Severe Weather Flying

AOPA/McGraw-Hill Series in General Aviation
General Editor: Robert B. Parke

HANSEN   Business Flying
NEWTON   Severe Weather Flying

Severe Weather Flying

Dennis W. Newton

036963

McGraw-Hill Book Company

New York St. Louis San Francisco Auckland
Bogotá Hamburg Johannesburg London
Madrid Mexico Montreal New Delhi
Panama Paris São Paulo Singapore
Sydney Tokyo Toronto

Library of Congress Cataloging in Publication Data
Newton, Dennis W.
  Severe weather flying.
  (AOPA/McGraw-Hill series in general aviation)
  Includes index.
  1. Meteorology in aeronautics. 2. Airplanes—
Piloting. 3. Thunderstorms. I. Title.
II. Series.
TL556.N48 1983   629.132'5214   82-7823
ISBN 0-07-046402-2   AACR2

1234567890   DOCDOC   898765432

ISBN 0-07-046402-2

*The editors for this book were Jeremy Robinson and Susan B. West, the designer was Naomi
Auerbach, and the production supervisor was Thomas G. Kowalczyk. It was set in Avanta by
ComCom.*

Printed and bound by R. R. Donnelley & Sons Company.

Portions of the Introduction and Chapters 1, 3, and 4 previously appeared in Dennis W.
Newton, "Thunderstorm!" *The AOPA Pilot*, June 1979; Chapters 5, 6, 9, and 10 in Dennis
W. Newton, "Thunderstorm!" *The AOPA Pilot*, July 1979; Chapters 12 to 14 in Dennis
W. Newton, "Icing Update," *Business and Commerical Aviation*, October 1978; and
Chapter 13 in Dennis W. Newton, "Icing Forecasts," *Business and Commercial Aviation*,
March 1979.

Contents

Foreword

Statistics show that forty percent of general aviation fatal accidents are related to weather. This suggests a real need for pilot education regarding the atmosphere. Proper weather decisions cannot be made without knowing weather. Knowing weather is not possible without understanding all of the elements and factors which contribute to its formation. Good flying and good decisions result from good pilot judgment which, in turn, results from having a fund of knowledge and experience from which to draw when faced with those continuing "go/no go" decisions. Pilots will find that reading *Severe Weather Flying* will add to their knowledge and help immensely in making those decisions.

No other subject area related to flying is as elusive as the subject of weather. Regardless of experience, pilots at all certification levels feel that their understanding of weather is incomplete. Because of the subtleties of meteorology, all of which interact with widely differing results, it is important to know all the machinery in "that big weather factory in the sky." Pilots need to build their understanding of weather on strong foundations, which will always remain in place ready to be used when needed. Otherwise, one is condemned to many frustrating and bewildering experiences when trying to go places in an airplane.

Few pilots are without at least a few troubling questions or misunderstandings about all that goes on around them in the sky. Dennis Newton's book dispels these mysteries of the atmosphere. You can fly more effectively and safely with the understanding that Newton brings to the subject than you can with a mystical concept of weather phenomena. Unlike writers of so many books in which an expert attempts to share his vast knowledge with his lay readers, Newton talks in plain language, pilot-to-pilot. As a teacher of weather, I view *Severe Weather Flying* as a great breakthrough in coming to know that medium of flight we call the atmosphere. Through that knowledge, pilots can maximize their use of the airplane as well as enhance the safety of flight.

Jack J. Eggspuehler
PRESIDENT
NATIONAL ASSOCIATION OF FLIGHT INSTRUCTORS

Preface

While I hope that practically anyone interested in weather, or in flying for that matter, will be able to read this book and gain some knowledge and enjoyment from it, it is primarily a book written for pilots by a pilot. As any prudent person would probably guess, it is mostly a book about how *not* to fly severe weather. Anyone except a knowledgeable research pilot with all the necessary safety nets who deliberately launches into a severe weather situation has pulled away from the factory a few bricks short of a load. (The cynics among you will probably already have guessed that the research pilot exception is there to cover my own adventures. All research pilots are crazy except me and thee, and sometimes I wonder about thee?) While occasional careless exceptions are not unheard of, pilots as a group are sane, responsible people who do not put themselves, their passengers, or their airplanes deliberately at risk.

Seekers after some advice on what to do if *caught* in a severe weather situation, a totally different matter which can and does happen occasionally, will (I hope) not be disappointed. However, to paraphrase an old saying in the flying business, superior weather pilots are those who demonstrate their superior judgment and knowledge of severe weather by avoiding situations which might force them to demonstrate their superior skill. I hope that this book will turn out to be a little help in the knowledge department. The judgment, of course, you'll have to supply for yourself.

In writing this book, I have found myself caught on the horns of an age-old dilemma—to wit, how to characterize weather elements such as thunderstorm structures and stability, which are so complex that they defy characterization, without resorting to mathematics or without so many except for's and whereas's that the thing degenerates into technical obscurities. Meteorologese is a language not readily translatable into everyday English. In addition, although my studies of meteorology have been extensive enough to earn a master's degree in the subject from Penn State, which definitely do not come in boxes of cereal, I make no claim to possession of the revealed Truth about weather or anything else. A lot of details will be revised as a result of both current and future research, and a lot of new lessons will be learned. However, if you let these obstacles be too much in your way, what you write is nothing, and I won't do that. I am aware of far too much aviation weather information which is not readily available to pilots for me not to try to get the word out where it can do some good. I have therefore taken the excellent advice of the ancient Roman philosopher who said that one should philosophize, but not too much. If one gets too nitpicking in presenting aviation weather—and particularly if one makes the all too common assumption (among weather guessers) that the pilot who doesn't happen to be interested in a particular esoteric detail is just not capable of comprehending it—then the only people who listen are other meteorologists. That amounts to preaching to the choir and ignoring the congregation, and I won't do that either.

"So get to the point," you might say. "What *will* you do?" Well, I'll do basically the same thing that scientists since Archimedes have done when confronted with a physical phenomenon too involved to be easily swallowed in one gulp. I'll resort to using more or less detailed conceptual models of the beast which will help to explain it. Such models stand or fall not on whether they are literally Truth in a philosophical sense but on how useful they are in making decisions about the real-world occurrences being modeled. If you understand the model, and if that understanding lets you draw correct conclusions about the "real thing," then the model has accomplished its purpose.

As a pilot who is writing for pilots (not as a meteorologist for whomever), when it comes to a choice of trying to draw a picture of the forest versus burying it in the trees, I will worry less about *true* in tedious detail than about *useful.* For example, in full knowledge that things are not really quite that straightforward, I will later make reference to the thunderstorm models developed by the late Dr. Fred Bates of St. Louis University. They are excellent tools in understanding the processes at work in the atmosphere and the hazards they present.

Will the details of any given thunderstorm differ from the model? Sure. The point is that the models are extremely useful from the pilot's viewpoint because they portray the hazards understandably and clearly enough to make them avoidable. That, after all, is the object. Likewise, pilots who savvy the explanations of stability in this book will not be able to whip out their pocket calculators and estimate Richardson numbers or equivalent potential temperatures (which terms, by the way, you have just seen for the last time in this book. Most such technical meteorological terms you will not find here at all). Does that mean that the subject has been oversimplified? I think not. Rather, I hope that it has been presented in a way which is appropriate to people who fly airplanes and that it will give them an insight into a whole big bag of weather tricks which they may not have had before.

This is a book not only about weather but also about airplanes and flying. While the emphasis is admittedly on types of weather which are potentially hazardous to flight, it is not a "cry-wolf." With each type of weather discussed, I have attempted to come up with rational answers to the pilot's very sensible question, So What? How much downdraft? How much wind shear? How much does the ice really weigh? Having done that, I have also taken a look at what the capabilities and limitations of airplanes and equipment are in avoiding and in dealing with severe weather. How much vertical gust is an airplane designed to take? What can really happen when an airplane takes a lightning strike? What does ice do to stall speeds? To drag? These questions are too often responded to with admonitions instead of answers, like "more than you can handle and it'll kill you dead as a doornail if you ever get near one of those things." That's what I mean by cry-wolf, and that's counterproductive because it's such an obvious exaggeration that it's widely ignored. Pilots don't have to be led around by the hand, and people who try to do that, however well-intentioned, at best accomplish nothing. At worst, they discredit themselves. In either case, when weather is presented in the worst possible light and it becomes the common experience of a pilot that there is no wolf, the weather briefer has no effective words of warning left when the day finally comes that the wolf is really there. Pilots as a group are more than conservative enough to keep themselves safe if they are only given the facts, and that's what I have done my best to do.

I have now reached the point where authors always say that their book would never have happened without the assistance of all sorts of people, and they proceed to name bunches of them. It almost embarrasses me not to be able to think of any better way of saying that, but I really can't because that's the way it is. I want to thank the Penn State University Department of Meteorology, particularly Dr. Dennis Thomson, for the opportunity of flying their research airplane for 2 really enjoyable years. This book was certainly born during that experience. Thanks also go to Bob Parke, Ed Tripp, Jack Olcott, and Archie Trammell for their help in publishing some of my previous work and encouragement in getting this book written. Andy Plumer supplied a helpful critique of the lightning chapter. Marion Maize took a thousand (more or less) words and turned them into a picture several times in providing most of the drawings. My appreciation goes to *Business and Commercial Aviation* and *The AOPA Pilot* magazines for their cooperation in allowing the use here of some material which previously appeared between their covers. Finally, a word of remembrance is in order to the late Dr. Fred Bates. Too many people knew us both for me to claim to have been fond of him, but that does not in any way diminish the credit due him for identifying a hazard and doing his best to get the word out to pilots. It is my opinion that the value of his severe thunderstorm concepts to safety in weather flying has been underappreciated for a long time.

Dennis W. Newton

Introduction

July 23, 1973, was not your normal day, at least not in St. Louis. I was the pilot of one of several weather research aircraft working a large urban weather project, and one of our jobs was to penetrate building storm systems as they moved toward the city. Although there were some storms reported in Missouri, and although every indication had been favorable, we had waited all day for something to happen in the St. Louis area with nothing to show for it but empty soda pop cans. Finally, in the late afternoon, we made one last radar check of the area. There was nothing on the scope but one small buildup to the northwest of the city, which had been sitting there doing nothing for some time. Tired of watching the proverbial pot, we scrubbed the cloud flight and proceeded to launch a low-level air-sampling flight crisscrossing the city VFR at about 1500 feet AGL. We fueled for a little over 2 hours of flying and bored off into one of the biggest weather surprises that I, for one, ever hope to have.

We were no sooner airborne from the Alton airport when, looking toward the northwest, we saw an ugly, dark-looking mass of cloud with a greenish cast and lightning all over the place. We flew up to the initial point for our air-sampling runs, which was about 10 miles north of the Alton airport, and we could see a tremendous, low roll cloud coming toward us with torrential rain behind it. We turned southwest and hit the gust front ahead of the storm. We began taking hard knocks, with everything not securely attached flying around in the cabin and downdrafts in excess of 1000 feet per minute. St. Louis approach control declared itself unable to handle VFR traffic at about that point, and there was no doubt about why. The entire city was solid thunderstorm, Lambert Field had gone IFR, scud was already visible at the river, and as I watched this display in amazement I was treated to a magnificent lightning stroke from cloud to ground to the east of Parks airport, about 25 miles south of my position and already well east of the Mississippi.

I ran. Alton was still open to the west, and I knew where the leading edge of the gust front was and elected to try it. Anyplace else would have meant driving off VFR with no plans and a totally blown weather picture and possibly having to leave the airplane parked outside in torrential rain, hail, or worse. We put it on the ground in a 4-knot wind which had become 29 knots by the time we cleared the runway. We were in the hangar just 30 minutes after we left it, happily thinking how much nicer it is to be on the ground wishing you were flying than vice versa.

Twenty minutes later, an Ozark Airlines FH-227 went down in the storm, on an instrument approach to one of the runways at Lambert.

What is this thing that sends an airplane full of unsuspecting meteorologists running for cover and then brings down an airliner? What causes it? Airplanes have flown through thunderstorms for years, practically since the development of gyro instruments, but once in a while one doesn't make it. Are these things flyable? If some are and some aren't, how do we tell which ones aren't? Airplanes have broken up in flight several miles from a storm, some of them apparently in clear air. Crashes have occurred on takeoff or landing as a result of thunderstorm wind effects several miles from the storm itself. Can this be anticipated and avoided? What about lightning? Is it dangerous to an airplane or not? There are answers, of sorts, to all these questions.

Some of them are good, some not so good, but all of them are a lot better than no answer at all. Some of them are obvious to anyone who gives the matter any real thought. Some of them, on the other hand, are subtle, and not at all what you would expect. Which are which? Is the obvious answer really the correct one? When I was a young charter pilot and flight instructor, I asked a pilot who was an old pro, and as weather wise as most pilots I knew, how he decided whether or not to fly a thunderstorm. He said, "Well, it more or less depends on how bad I want to get where I'm going." I didn't find that much of an answer and sort of felt put off at the time—you know, "go 'way kid, ya bother me." I later realized though that he was just leveling with me and not trying to snow me. He really didn't have an answer. We can do better than that now.

List of Abbreviations

ADF	automatic direction finder
AFOS	Automation of Field Operations and Services
AGL	above ground level
ATC	air traffic control
Cb	cumulonimbus cloud
CRT	cathode ray tube
Cu	cumulus cloud
FAA	Federal Aviation Administration
FAR	federal aviation regulations
GS	groundspeed
IAS	indicated airspeed
IFR	instrument flight rules
MSL	(above) mean sea level
NACA	National Advisory Committee for Aeronautics
NASA	National Aeronautics and Space Administration
NTSB	National Transportation Safety Board
OAT	outside air temperature
PIREP	pilot weather report
SEA-TAC	Seattle-Tacoma
SIGMETs	significant meteorological condition forecasts
VFR	visual flight rules
VORTACs	very high frequency omnidirectional radio range/tactical air navigation station

1
The Four Fundamentals

Four basic ingredients go into the recipe for severe weather, and essentially all other weather for that matter. They are as follows:

1. *Water.* All weather (except some of the winds which serve to move the water around) is made of water. Icing clouds are made of water. Thunderstorms are made of lots of water. Water, of course, is everywhere. About two-thirds of the planet is covered with it. Water to make weather out of enters the air from oceans, lakes, and rivers. It is no coincidence that some of the most treacherous weather in the world occurs in the area of the Great Lakes. Enough water to make a lot of weather can even come right off the ground, particularly after a rain. Everyone has seen a day begin to dawn bright and clear after a night rain, only to sock in tight as the morning heating lifts water into the air. When you look at any sort of weather chart, ask yourself where the water is. What are the dew points? Are the winds coming from dry land or from a source of moisture? Water is *the enemy.*

2. *Temperature.* Various types of weather require temperatures which are somewhat, but not too much, below freezing. Fog requires temperatures near the dew point. Thunderstorms require relatively warm temperatures in the lower layers of air in which they form for the simple reason that warm air can hold more water.

3. *Lifting.*   Very early on, you are likely to find in most basic weather texts for pilots some statements to the effect that low-pressure areas are associated with obnoxious weather. Why should this be true, you might ask yourself, looking probably pretty much in vain for an explanation. It's really quite simple. Air is drawn into the low near the surface. It can't pile up there. Where does it go? Up. *Voilà!* If moisture is present in sufficient quantity, the result is weather. It is a small oversimplification indeed to say that all a front does is lift air. There are such things as sea breeze fronts and dew point fronts, in addition to the commonly known cold, warm, and occluded fronts, which are in the business of lifting air. The jet stream, and other smaller-scale upper-air wind flows, creates "holes" in the upper air that result in lifting of low-level air to fill them. Hills lift air. A thermal over a hot parking lot surface lifts air.

4. *Stability.*   Stability is one of the most important factors in weather, and one of the least understood. If you want to understand thunderstorms, icing, or any other kind of weather for that matter, you have to understand stability. Fear not. Stability is very simple and even fun—it is only the explanations that are weird and mysterious. More on this subject is forthcoming a bit later.

As a large-scale example of the effects of the Four Fundamentals, wouldn't we expect to find that more thunderstorms occur in areas where there is lots of warm water and something to lift it than in other places? Regard the southeastern United States (see Figure 1-1). The Gulf of Mexico is one of the best warm-water sources in the world. We have hot land and sea breeze fronts to create lifting in Florida—an area which is surrounded by warm water and apparently has the most thunderstorm days (days on which one or more thunderstorms are observed by the National Weather Service reporting network) in the country. I say "apparently" because there are also lots of stations in Florida to report thunderstorms, and that may have a little to do with

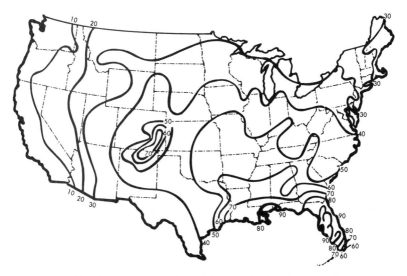

FIGURE 1-1 *Yearly average number of days with thunderstorms based on observations from all U.S. Weather Bureau first-order stations in the United States. A thunderstorm day is considered any day during which one or more thunderstorms occur. (FAA.)*

FIGURE 1-2 *Orographic thunderstorms building in Colorado.*

it. Look at the mountainous area in Colorado and New Mexico. Warm, moist air from the Gulf flows in there, too. Not as much as in Florida or the rest of the southeast, probably, but there is lots of lifting due to the terrain which will continue to try to squeeze the water out. [The area to the west, into Utah, could easily be a little more active than is indicated—the low numbers shown could be due to the low density of reporting stations in the area. I've sure seen enough storms (see Figure 1–2) while flying in that area.] Two things which are apparent from Figure 1-1 are the deep penetration of Gulf air into the United States and the effect of lifting by terrain.

So much for an introduction to the Four Fundamentals. They are basic to everything from here on, and they make the understanding of weather a lot easier. Two of them, lifting and stability, are covered in more detail in the next two chapters.

2 The Ups and Downs of Air

The atmosphere can be thought of as a heat engine, driven by the sun, with water as the working fluid which transports energy. Lifting of air which contains water is a prerequisite to the creation of most weather which is potentially hazardous to flight. It is therefore worth a little time to think about the mechanisms in the atmosphere, both large scale and small scale, which do this lifting.

On a nonrotating earth, there would be no large-scale lifting mechanisms, e.g., fronts, in the mid-latitudes. Heated air would rise in the vicinity of the equator. Cold air would descend near the poles. The result would be a single large-scale cell of circulation in each hemisphere. In the northern hemisphere, this would be southerly winds aloft transporting air toward the pole and northerly winds near the surface carrying it back toward the equator. In the real world, things are not that simple. There are actually three cells of circulation in each hemisphere, instead of one. Rising air near the equator descends in the subtropical latitudes. Descending air near the pole rises in the mid-latitudes. The third cell fits between these two, as shown (see Figure 2-1). "Aha," you say, looking at the area where the middle cell and the northern cell meet, "lifted air!" Right you are.

This area of lift, where the polar air meets the mid-latitude

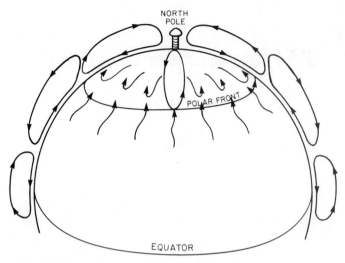

FIGURE 2-1   *A simplified sketch of atmospheric circulation. Note the lifting at the Polar Front.*

air, is the Polar Front. If nothing else were going on, this front would more or less sit there, sort of like a bathtub ring of crumby weather around the earth, maybe moving a little south as a cold front and a little north as a warm front now and then. Naturally, though, something else is going on.

First there's the Coriolis force. Without belaboring the physics, the Coriolis force is a phenomenon caused by the rotation of the earth, which makes everything in motion want to turn to the right. It's pretty small, but then molecules of air are pretty light. Since air in the middle cell of circulation is northbound at the surface, the right-turning effect results in a belt of mid-latitude westerlies. Low-level easterlies, of course, are the result in the northern and southern latitudes. (By the way, for the benefit of any sceptics to whom the Coriolis force sounds like black magic, it can be observed in much larger things than air, given enough time. Rivers erode the right bank. The right rail wears out first in tracks traveled only in one direction.)

Second, there's the jet stream. A complicated result of the atmospheric circulation pattern is the apperance of the westerly jet stream in the upper air above the Polar Front. What is the jet? By arbitrary definition, a narrow band of upper winds is called a "jet stream" when its speed exceeds 50 knots. However, it's common knowledge these days that the wind speeds in the core of the jet are much higher than 50 knots, easily triple that at times. Consider, for a moment, what that means. At those speeds, airflow is basically incompressible. The high speed in the core of the jet then either means a much lower pressure than at its slower-moving west end, or that a lot more air is moving in the middle than at the end, or both. Either way, the conclusion is that the jet core acts like a big vacuum sweeper. It wants air, more air! Where does the air come from? The best source of air is beneath it. So, a surface low-pressure center is induced in the vicinity of the Polar Front by the jet stream aloft, and now all the ingredients of the large-scale lifting mechanisms are in place (see Figure 2-2).

Air at the surface starts moving toward the low center created by the jet on or near the Polar Front. Coriolis says that the air

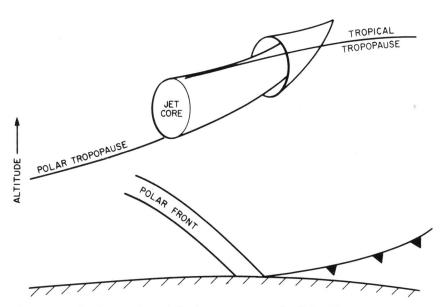

FIGURE 2-2 *A cross section of the jet stream over the Polar Front.*

has to turn right. It does, and now we have a counterclockwise circulation around the low. This starts the Polar Front, or pieces of it, rotating around the low. The jet stream suction moves generally from west to east and tows the low and what has now become an associated cold front, warm front, and possible occluded front, along with it, and there we have it. The large-scale lifters are in full operation. There is general lifting in the low-pressure system as air is drawn upward by the vacuum sweeper. Air coming from behind what is now a warm front is driven up over it, adding to the general lifting. Further behind, the now moving cold front is plowing underneath the air in front of it like a giant cow catcher, creating even more energetic lifting, perhaps catching up with the warm front and adding to its lifting effect in an occlusion (see Figure 2-3).

Ok, so now we have built up a picture of what a mid-latitude low-pressure system is, we have seen what drives it, and we can begin to visualize where the weather will be found in terms of where the lift is. Does that mean that without this large-scale lifting there would be no weather? Common experience says that the answer to that question is no, since severe weather, and a lot of other kinds of weather, occur well away from any low-pressure system. You could then conclude that there must be other smaller-scale lifting mechanisms, and you'd be absolutely right. On

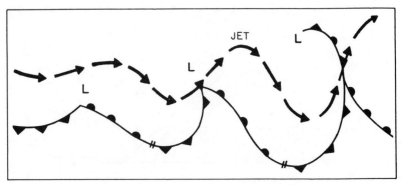

FIGURE 2-3 *Average jet stream positions over surface frontal systems. (FAA.)*

a very small scale, a hill with an upslope wind lifts air. Hot ground starts thermals going which lift air.

On a somewhat larger scale, there are vacuum sweepers in the upper air called "troughs" which can kick off lifting not associated with surface low-pressure systems. More down to earth, there is such a thing as a dew point front. One such dew point front is sometimes called the "Marfa front" because it is often found in the vicinity of Marfa, Texas. It travels widely, however, and at any given time it could be anywhere from eastern Colorado and New Mexico to Missouri and eastern Texas. Moist air from the Gulf of Mexico has to meet dry Rocky Mountain air somewhere, and the Marfa front is where it does it.

Dry air being more dense than moist air, the dry air will push under the moist air at the dew point front and provide lifting much like a cold front even though there may be no appreciable temperature difference across the front. The Marfa front frequently kicks off squall lines when it migrates eastward lifting the moist Gulf air ahead of it. Such a dew point front also frequently occurs 100 miles or so ahead of a fast-moving cold front. This may be due to air descending from aloft and consequently drying

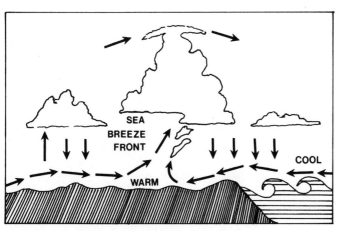

FIGURE 2-4 *A daytime sea breeze front. (FAA.)*

FIGURE 2-5a to c *Lifting just onshore on a September day in Florida kicks off cumulus development which quickly builds into thunderstorms.*

FIGURE 2-5b

FIGURE 2-5c

out just ahead of the front, but whatever the reason, the result
is often a squall line well ahead of the actual cold front. Squall
lines can be very misleading, since they have a lot of the symp-
toms of a normal frontal passage and can sucker you into believ-
ing that the cold front has passed when it is actually still lurking
to the west.

One other small-scale lifter is the sea breeze front (see Figure
2-4). During the day, a circulation often develops from cool water
to warm land at a shoreline which lifts air like a miniature cold
front. At night, on the other hand, the circulation can develop

in the other direction, with cool air from the land lifting air over the water. Either can trigger thunderstorms if conditions in the air being lifted are favorable (see Figure 2-5a to c). The ones at night, particularly, can be a nasty surprise to the unwary.

So much, for the moment, about lifting. The more air raising that goes on, the more hair-raising the resulting weather is likely to be. Still, lift is only the trigger. No water, no weather. If there is water to lift, then stability really gets into the act, and that's the next subject.

3 *Stability*

When we talk about something being stable or unstable, what we are discussing is how it reacts to being disturbed. This is true whether we are referring to an airplane, a personality, an air mass, or anything else which can be momentarily upset and then set free to do its thing. Something which is stable will return to its original course after the disturbance goes away. Thus, your St. Bernard scratching in the baggage area will do no permanent harm to the flight path of your Stinson, because it's stable. An unstable situation, on the other hand, results in the thing in question flipping out after it is disturbed and either never returning to its original situation or returning at such a fast rate that it rockets through and goes merrily off in the opposite direction.

We need to know two things to understand stability as it relates to the atmosphere. Are you ready?

1. Hot air rises. (Also, to no one's surprise, cool air descends.)

2. The sun does not heat the air to any great degree. The sun heats the surface of the earth, and the land and water heat and cool the air.

Now, knowing these two things, let's see what we can figure out. How about this? Warm over cold is stable. Hot air rises, right? Then warm over cold is the way things "druther" be, if they had their druthers. When we say warm and cold, of course, we are talking in relative terms. If the temperature we see on our airplane thermometer rises as we climb, or even if it just remains the same and does not decrease, we have a case of warm air over cold air, and that's stable. On the other hand, if the temperature drops rapidly as we climb, we have a case of cold over warm. This is against the nature of things. The warm air at the bottom wants to rise, and the cold air above wants to descend. Give it a push, and this layer of air will literally overturn. Make it hot enough at the bottom and it will overturn even without a push.

Suppose the air does overturn. A layer of air obviously can't roll over like a hibernating bear. What it does is start at the hot spots, or in areas where it is given a push. We then have an area in which the warm air is rising. The atmosphere doesn't like holes, so more low-level air moves into these areas and then it, too, rises. The cooler air above descends to replace the air which is leaving the lower levels.

This process will shortly result in a stable, warm over cold, situation, which is what the atmosphere likes. Unless, of course, the reason we had a cold over warm case in the first place was because the ground is hot. If that's the case, then the cool air which came down shortly becomes hot air, and we have a steady-state, unstable situation.

Everything that's been said about stability up to now is true whether or not the air is moist. Now suppose the air that's going up contains water vapor. The warmer the air is, the more water vapor it can hold. However, if you begin to cool the air, sooner or later you reach the point, called the "dew point," where the air holds all the water vapor it can. At that point, if you cool the air any further, a cloud appears, composed of small droplets of liquid water. We can make a pretty good guess at where this will happen. As a bubble of dry air (that is, air which may contain water vapor but in which no cloud exists) rises in our unstable air mass, it cools at a rate of 5.5°F for each 1000 feet that it rises. The surface dew point which our bubble of air had decreases at the rate of about 1°F for each 1000 feet of ascent. Consequently, the temperature of our bubble, which is decreasing as it rises, is approaching the dew point, which is also decreasing slowly, at the rate of 4.5°F for each 1000 feet of ascent (see Figure 3-1). To estimate the cloud base, then, we take the difference between the surface temperature and the surface dew point and divide by 4.5. For example, if the temperature at the surface is 68°F and the dew point is 50°F, and if the situation is unstable and air begins to rise, clouds will begin to form with bases at about 4000 feet AGL.

When water vapor condenses into liquid droplets, it releases heat into the air which contains it. Consequently, if our bubble of air (which now contains a cloud and is receiving heat from

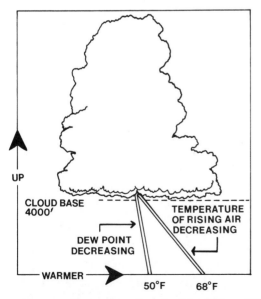

FIGURE 3-1 *Formation of a cloud by lifting. (FAA.)*

the water) continues to rise, it will cool off more slowly. The rate varies, but it will be something in the neighborhood of 2°F per 1000 feet of ascent. Now, if we are interested in the possibility of thunderstorm formation, we must ask ourselves this question. *Will the air in our rising bubble become warmer than the air which surrounds it?*

First of all, suppose we have a case of warm over cold in the layer of air in which our bubble starts to rise. Then the answer to our key question is *always no.* We have already said that warm over cold is stable. Since our rising bubble is cooling, quickly if there is no cloud forming or more slowly otherwise, but cooling nonetheless, it will always be colder than the air around it. Cool air descends, so our bubble will head back whence it came, and that, after all, is what we said we meant by stability.

Now, suppose that the layer in which our bubble of air starts to rise is a layer of cold air over warm. Suppose, as you read the thermometer in your airplane climbing through this layer, you see that the temperature drops at a rate of more than 5.5°F per 1000 feet. In this case, the answer to our key question is *always yes.* The bubble can't cool any faster than 5.5°F per 1000 feet, whether or not there is a cloud forming. Therefore, as soon as you lift it a little, it will always be warmer than the air around it. We have said that a layer of air which is cold over warm is unstable, and this is why.

The reason why most thunderstorms occur during the afternoon is now as plain as the difference between night and day. Just exactly as plain, in fact. During the night, the ground cools off and generally results in a layer of warm air over cold. This, of course, is stable. During the afternoon, on the other hand, the ground is hot, and an unstable, cold over warm, situation results. The temperature in a layer of air over hot sand can be 50° colder 10 feet off the ground than it is at the surface. I have flown aircraft with infrared-temperature-sensing equipment over large areas of asphalt in the afternoon and found the asphalt temperature near 150°F, about 70° warmer than its surroundings.

There is still one other possibility to think about. Suppose the temperature decreases with height as we climb and watch our thermometer at some intermediate rate, say 3°F per 1000 feet or so. Let's ask our key question again. Will the air in our rising bubble become warmer than the air which surrounds it? Now we get an answer which may be the most important one we have. *It depends on the moisture.* If no clouds are forming, our bubble cools rapidly, becomes colder than its surroundings, and settles back down. However, suppose a cloud forms as the air bubble rises. We saw earlier that the closer the temperature and dew point are at the surface, the lower the cloud base will be. Once the cloud starts to form, releasing heat into the air, the rising bubble cools off much more slowly. Now it stands a good chance of becoming warmer than its environment as it ascends. As soon as that happens, the brakes are off. The higher it gets, the warmer it becomes relative to the air around it, and the faster it moves. Perhaps it won't become a thunderstorm. It may hit a higher layer of stable air which will stop it, or it may run out of moisture. However, once our bubble has a cloud forming in it and becomes warmer than the air around it as it climbs, it sure is on its way.

Let's summarize what we know about stability:

1. A layer of warm over cold is stable. If the temperature of the air through which we are climbing increases, or at least does not decrease, convection, the overturning of air, will not start in that layer (see Figure 3-2).

2. A layer of cold over warm is unstable. The degree of instability depends on how fast the temperature decreases with height. A layer of air near the ground on a hot afternoon is likely to be very unstable and will start overturning at the slightest push whether moisture is present or not.

3. Intermediate between these extremes, the stability depends on the moisture in the layer. The higher the humidity, the lower will be the base of any clouds which form, and the more likely it is that the situation will result in fireworks.

The National Weather Service publishes a much underrated weather chart depicting stability, which is a grand place to start a weather briefing on what may be a thunderstorm day (see Figure 3-3). The lifted index, shown on the top of the horizontal line at each station on the chart, is the temperature difference which a bubble of air would have if lifted from the surface to the 500-millibar pressure level, which is a height of about 18,000 feet. The difference is in degrees centigrade, and a negative number means the bubble would be warmer than its surroundings if lifted. (The minus sign comes from the way they do the subtraction.) Therefore, negative lifted index values indicate instability. The so-called K index, underneath the line at each station, is more complex, but a high value of this number indicates a high moisture content. Therefore, this single, very useful little chart provides at a glance the big picture of areas in which thunderstorms are probable. A value of the lifted index of −2 (remember, this means that a bubble of air, if lifted, would be 2°C warmer than the air around it) is unstable enough to make tornadoes possible. A meteorologist named Modahl did a study of fifty-five general aviation thunderstorm accidents and found that the average lifted index existing at the time of the accidents was −4. The average lifted index in eighteen airline accidents which he also studied was a whopping −7.

FIGURE 3-2 *A classical case of stable air. Looking west into the Los Angeles basin, we see the smog in the trap.*

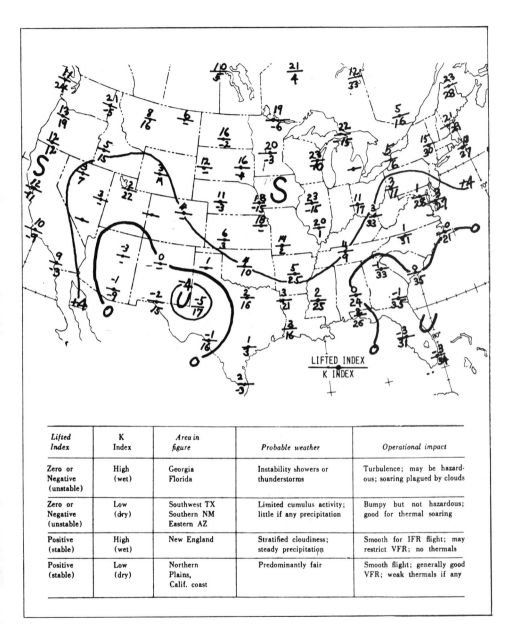

Lifted Index	K Index	Area in figure	Probable weather	Operational impact
Zero or Negative (unstable)	High (wet)	Georgia Florida	Instability showers or thunderstorms	Turbulence; may be hazardous; soaring plagued by clouds
Zero or Negative (unstable)	Low (dry)	Southwest TX Southern NM Eastern AZ	Limited cumulus activity; little if any precipitation	Bumpy but not hazardous; good for thermal soaring
Positive (stable)	High (wet)	New England	Stratified cloudiness; steady precipitation	Smooth for IFR flight; may restrict VFR; no thermals
Positive (stable)	Low (dry)	Northern Plains, Calif. coast	Predominantly fair	Smooth flight; generally good VFR; weak thermals if any

FIGURE 3-3 *Stability chart. (FAA.)*

You may have noticed that in all the discussion up to now there has not been a single mention of the so-called standard lapse rate, which is roughly 3.5°F per 1000 feet. There is a reason for that—to wit, the standard lapse rate doesn't have a thing to do with weather. If you believe that sacred cow makes great hamburger, this particular one is long overdue for the grinder. The standard lapse rate is an engineering standard used for calibrating altimeters, and not much else. As we'll see a little farther on, blind use of the standard lapse rate to estimate temperatures aloft can lead a pilot into gross tactical blunders, like descending into an icing condition on the mistaken assumption that it will be warmer down there than it actually turns out to be. The standard lapse rate is science fiction.

4 *Air-Mass Thunderstorms*

Flights which have to deal with air-mass thunderstorms are common during much of the year in the United States. Take, for example, the northeast or southeast on a late spring or summer day under the influence of a stagnant, fairly humid high-pressure area. No fronts, not too much low-level wind, visibility at the surface reported as probably 10 miles or less. Perfect for air-mass storms. The first thing we will do, naturally, is check weather. If thunderstorms are forecast along the route, we will make note of where and when. We always take these forecasts with a grain of salt, however. If no thunderstorms are forecast, that does not necessarily mean that none will occur. If thunderstorms are expected, and if they do occur, they may not happen when or where expected. The distance between primary weather stations is about 100 miles, on the average, and the distance between upper-air stations is more like 200 miles. Surface observations are made hourly but usually only every 12 hours at the upper-air stations. This essentially limits forecasting to a statement that storms are probable in an area much larger than the storms themselves, and while such forecasts are far from useless, there is no way of knowing in advance that one will develop on Victor 16 or over the East Filchboro Airport on a particular afternoon. If we don't have radar or a Stormscope, we must use our eyes and what we know about moisture, stability, temperature, and lifting to fill in the gaps.

If possible, we plan to go early in the day. The air is generally more stable at that time, and hot spots on the ground to start thermals are fewer. When we get our weather briefing, we will ask where and when thunderstorms formed along our route on the previous day. If the weather situation is about the same as yesterday, the thunderstorm situation probably will be too. If we can, we will also find out where the top of the haze layer is and plan to fly above it. That will allow us to see the action when it starts. We will peer from our window in the morning, and we will check current weather along our proposed route even if we are going to depart several hours later, looking for cumulus and altocumulus clouds. When the low-level Cu starts to bump, we know that instability is present. The earlier in the day and the lower the cloud base the more chance there is that some of that Cu will become a Cb, the cumulonimbus cloud that we don't want to tangle with (see Figure 4-1). If we see altocumulus, the cumulus cloud that forms at around 12,000 to 16,000 feet above ground, we know that there is a moist and unstable layer aloft. This, by itself, does not indicate thunderstorm formation, but it tips us off that any low-level cumulus clouds which get that high will find a good environment to continue growing. A towering cumulus cloud which isn't doing much and looks like it's about to poop out can turn into a real tiger if it penetrates such a midlevel, unstable layer, and if we realize the layer is there, we won't be taken by surprise. If we have a moist wind in the lower 5000 feet or so, we will look for a route which keeps us well upwind of rising terrain, or well away from it in some other direction if upwind is not possible, because we know that the upslope wind may provide the lift necessary to trigger orographic (upslope) storms. If we have to cross a mountain range, the earlier in the day and the lower the terrain where the route crosses, the better are our chances of being where the storms ain't. Then, having prepared ourselves as best we can with forecasts, a good route,

and what we can see for ourselves, we make one last call to Flight
Service to check the latest weather for the existence of thunderstorms along the route, we mount our aeronautical steed, and
we're off.

If the day permits it, we climb to the top of the haze layer,
level off in the clear, and look around. We scan the horizon in
the direction we're headed for indications of dense cirrus, which
may indicate a thunderstorm top. Wisps don't bother us, but if
the horizon appears overcast with cirrus or if we see high clouds
which become thick in the upwind direction, we may be looking
at a thunderstorm anvil. This is our earliest visual warning that

FIGURE 4-1 *A morning altocumulus layer in Florida.*

FIGURE 4-2a to c *Cumulus developing and shearing off. No thunderstorm danger here. These pictures were taken 2 minutes apart.*

FIGURE 4-2b

FIGURE 4-2c

actual thunderstorms exist and can often be seen when we are
still 200 miles or more from the active storm cells. We will keep
an eye on such areas as we move along, watching whether they
develop or move. If a route change appears wise, we can probably
make it early in the game so that it will only have to be a few
degrees and won't affect our flying time much. We look around
for cumulus popping up through the top of the haze, and when
we see some we spend a little time watching it. If the winds are
right and there is not too much moisture, the tops of the clouds
may shear off and dissipate in 5 minutes or so (see Figure 4-2*a*
to *c*). Areas where that happens are fat city, for the time being
at least, and we can rejoice. Look up ahead there, though, at
about eleven o'clock. See that one popping up? It's standing up
straight, and the top looks hard, like a cauliflower, not fuzzy
around the edges. We watch, and 15 minutes later there's no
doubt about it. The cloud towers into the sky and the cirrus is
spreading from the top into the characteristic anvil. It's a thunderstorm, all right. Well, it's not in our way, but let's watch. Sure

enough, there's another Cu popping up upwind (see Figure 4-3 *a* to *c*). This is what we're looking for. Thunderstorms tend to form in lines, more or less upwind and downwind. Forty minutes after the first sighting of a building Cu, we have three full-blown thunderstorms getting lined up to do battle. Now we know where the activity is, and since our destination is on the other side of the forming line, we get set for an end run. The line may close up, and we want to be on the other side if and when that happens. The Cu is building at the point we pick for crossing when we get there, but the storms are not connected. We pick a spot where we can see lots of sky on the other side of the activity and the clouds are not solid above us. We examine the building cloud below us, and find an area where the tops are fuzzy and not active (see Figure 4-4*a* to *c*). The hole can close in a hurry if we try to penetrate over a sharply defined, cauliflowery, building Cu. In about 5 minutes we're home free, back on course to our destination with the activity behind us.

This is usually the best way to handle an air-mass-storm situation. If there is a lower cloud deck which prohibits a VFR climb to an altitude in the clear, the pilot who is instrument-rated is generally better off to get on top of the lower deck and then visually avoid the storms. If you are not instrument-rated or if you are and can't get clear of cloud, the problem is more difficult. You will have less warning and should make frequent weather checks and obtain whatever radar assistance may be available from the air route traffic control center or approach control facility serving the area. If you don't have instrument enroute charts, borrow them from someone long enough to note the center frequencies along the route. Even if the ATC radar is not painting the weather, just listening in to the conversations of other aircraft in the area, some of which will be flying in the clear, will help you locate the activity. However, you will probably not have any early indication of which way the storms may be lining up. The pilot without thunderstorm detection equipment, tooling along in the murk with 5 miles or so visibility and unable to climb above

FIGURE 4-3a to c *Air-mass thunderstorm development. This development occurred in 35 minutes.*

FIGURE 4-3b

FIGURE 4-3c

FIGURE 4-4a to c   *VFR through the line and into the clear before things close up.*

FIGURE 4-4b

FIGURE 4-4c

it, must be ready and willing to turn back and land if necessary. When the sky becomes dark ahead or it looks like you are flying into a wall (which will shortly prove to be water), or you see lightning or turn the squelch on the radio up and hear the crackling that sometimes indicates lightning, it's time to quit.

If you are instrument-rated and are flying without radar or a Stormscope, and if you fly long enough in IFR conditions favorable to air-mass thunderstorms, sooner or later you are going to get caught. It happened to me on a trip home from a cloud project in St. Louis. The Stormscope hadn't been invented yet and we had sacrificed airborne radar for a gust probe research installation in the nose. New York Center wasn't painting the weather that day, and after about 95 hours of deliberately flying clouds with guidance from our own ground-based radar and never really getting into anything I didn't want to tangle with, there I was, on what was supposed to be a routine trip, on Victor 30 at 7000 feet (more or less) in a thunderstorm.

Which brings us to the question, Are these things flyable? You can have your choice of answers with respect to air-mass storms.

"Yes, but" or, "No, but." A true air-mass thunderstorm is the kind we have all known about for years (see Figure 4-5). It begins with the cumulus (building) stage, characterized by building cloud and updraft. How much updraft? I have seen 3000 feet per minute any number of times. Ask yourself how fast a storm can get to 30,000 feet at that rate and you can see how we had three full-blown storms in sight in 40 minutes on the armchair flight we made earlier. The air-mass thunderstorm stands erect in a generally fairly light wind field and is said to reach the mature stage when rain begins to fall. Since the storm is essentially vertical, the rain falls down through the updraft. This causes two things to happen. First, the drops have aerodynamic drag, just like any other object moving through air. This creates downdrafts. The water hits the ground and sticks, more or less, but the downward rushing air spreads out away from the storm, mostly ahead of it, creating a gust front to amuse the unwary aeronaut who tries to take off or land too close to the storm (see Figure 4-6). How much downdraft? Since airplanes also tend to stick when they hit the ground, I don't fly into rain under thunderstorms at low altitudes to find out. However, I'd believe 2500

FIGURE 4-5 *The stages of an air-mass thunderstorm. (A) is the cumulus stage, (B) the mature stage, and (C) the dissipating stage. Arrows depict airflow. (FAA.)*

feet per minute easily. Never take off or land in the face of an approaching thunderstorm, and particularly not into its rain.

The second thing that happens due to water falling through the updraft is that the storm strangles. The air-mass storm is, therefore, self-destructive, with a life cycle of usually 20 to 90 minutes. This kind of storm does not generally contain hail. Numerous penetrations of such storms were made during a thunderstorm study conducted in Florida and Ohio during the late forties. Ninety percent of the penetrations flown at 16,000 feet and 100 percent of penetrations at 6000 feet found no hail in Florida, and in Ohio the numbers were 94 percent no hail at 20,000 feet and 98 percent at 5000 feet.

Icing in any thunderstorm can obviously be ferocious above the freezing level. The updraft regions are by far the worst areas, because that is where water vapor is condensing into liquid drops the fastest. About −10°C is usually considered to be the lower limit of temperature for serious icing, but that's not true in a thunderstorm updraft. Liquid water is condensing in these drafts at a rate much faster than it can freeze in the cloud, and literally

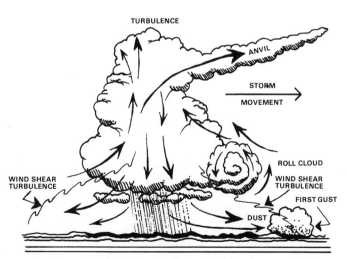

FIGURE 4-6 *A simple air-mass thunderstorm and gust front. (FAA.)*

no altitude above the freezing level can be considered safe from severe icing in a Cb updraft.

The air-mass storm (by which we mean the type of thunderstorm just described, not necessarily any storm which forms without frontal or orographic lifting) is the gentlest kind of thunderstorm, and we have seen that it is more than sufficiently dangerous to warrant strict avoidance. Although the vast majority of such storms do not contain weather which is destructive of an airframe, there is just no way of being certain that any particular storm does not contain weather that can literally swat you out of the sky. There are basically two other kinds of thunderstorms, both of which contain weather at least damaging if not catastrophic, and there is no way of guaranteeing that the storm you are looking at is not, or will not suddenly turn into, one of these other kinds. We will call the other two kinds the "steady-state storm" and the "severe storm." That's more or less in accordance with the general usage of these terms, but take a little care in reading other things on the subject. The definitions may be a little different. We will have a look at these other types of thunderstorms shortly. But first, let's go back to basics for a moment.

The basics are four things: water, temperature, lifting, and stability. Water provides the energy needed for thunderstorms, and the presence of sufficient moisture is generally indicated by the dew points found in the hourly sequence reports. A dew point of 53°F or greater is enough moisture for very severe thunderstorms, possibly with tornadoes. There is no such thing, however, as a light thunderstorm. An ordinary air-mass storm has the energy equivalent to about thirteen World War II atomic bombs.

Temperature and moisture are related, as are temperature and stability. The air has to be warm enough in the lower levels to hold enough water vapor to fuel the storm. The instability necessary to get thunderstorms going is the result of layers of air in which the temperature decreases with height, cold over warm. Moisture and stability are related. The more moisture in the lower levels of the atmosphere, the lower the cloud base will be and the more likely that a rising bubble will become warmer than the air which surrounds it and head for the stratosphere.

Lifting is what sets it off, the match that lights the fuse of this atmospheric firecracker. If a moist parcel of air is driven up a hill so that its temperature reaches its dew point and it becomes warmer than the air around it, we have the beginning of a thunderstorm (see Figure 4-7). A hot parking lot on a summer afternoon can provide the thermal lift necessary to get things going. Any kind of front can lift air along a line hundreds of miles in length. Combine frontal lifting with lots of moisture and high instability, and we are talking about storms as destructive of property on the ground as an artillery barrage, and as destructive of aircraft as a Sidewinder missile. Storms with this destructive capacity can occur along with ordinary air-mass storms, and we will ordinarily not be able to tell them apart by looking at them (see Figure 4-8). One reason that it is worthwhile to know the lifted

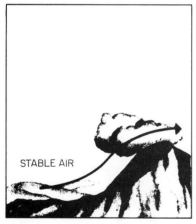

STABLE AIR   UNSTABLE AIR

FIGURE 4-7 *When stable air (left) is forced upward, the air tends to retain horizontal flow, and any cloudiness is flat and stratified. When unstable air (right) is forced upward, the disturbance grows, and any resulting cloudiness shows extensive vertical development. (FAA.)*

FIGURE 4-8 *An air-mass thunderstorm building near Allentown, Pennsylvania, on a warm, humid August afternoon. The middle clouds present are being blown off of other storms and will soon make visual storm avoidance at low altitude impossible.*

index from the stability chart is that it allows us to assess the potential for destructive storms. Remember, fifty-five general aviation aircraft that we know of have been swatted out of the sky by storms when the lifted index was around −4.

Armed with our knowledge of the basics and of the air-mass storm, we are ready to take a look at the Mama Bear and Papa Bear of the thunderstorm family—the steady-state storm and the severe storm.

5 *In This Corner, Mama Bear*

The air-mass thunderstorm, which we have discussed earlier, is the three-stage thunderstorm—Baby Bear, if you will. It builds, rain starts to fall as it matures, the rain strangles the updraft, and the storm dissipates. To get a steady-state thunderstorm, we have to take the brakes off. We must get the water out of the updraft.

The simple way for this to occur is to have the storm develop in an environment in which the wind changes with height. This usually means that the wind increases with increasing altitude. The more the growing thunderstorm leans, the more water will go elsewhere than right back down through the updraft, and the more the storm can grow. The thunderstorm which slopes can produce hail in large quantities and large chunks. I was first made aware of this by the late Dr. Fred Bates of St. Louis University, and I found it useful as an Air Force forecaster in New Mexico in the mid-sixties. When I observed a thunderstorm with the radar, I would elevate the antenna and scan it vertically. If it sloped, I issued a hail warning. In many cases, of course, the storms didn't pass over the base, so I have no way of knowing if hail fell out of them. Those which did, however, usually hailed. Remember, though, that I was looking for a sloping updraft and using a radar set which painted liquid drops, which were being generated copiously by condensation in the rising air. You may be able to tell that some given storm slopes with the old eyeballs. You also may not be able to, since there may be cloud all over the place. The fact that you don't see slope doesn't mean it isn't there, or that hail is not present.

I mentioned that I found this sloping-storm idea useful in New Mexico. Why there, particularly? Well, where would we expect to find a situation where the winds frequently increase with height? How about mountainous terrain, where low-level winds are interfered with by rockpile? As we have seen, the maximum number of days per year with thunderstorms occurs in Florida, but hail at the surface is nearly nonexistent there. The maximum frequencies of hail days occur on the eastern slopes of mountain ranges in Colorado, Wyoming, Nevada, and Utah, in areas getting around half the number of thunderstorm days as Florida. Colorado, of course, is next door to New Mexico, which is also mountainous. Any thunderstorm forming in such terrain should be suspected of being a hail producer, and we don't have to penetrate the storm to run into it (see Figure 5-1). Shafts of hail may fall from high in the storm to its rear and from the anvil downwind of the storm. The only safe course of action is to give the storm a wide berth. The hail could be 10 miles or so from the rain underneath the anvil. A hail shaft often has a greenish appearance and may sometimes be sighted visually. The fact that you don't see one, however, doesn't necessary mean it isn't there or won't appear in your location if you circumnavigate only the rain. I know of two separate incidents in which a research airplane penetrated a thunderstorm without encountering hail but another closely following the research airplane hit a hail shaft, in both cases inadvertently.

Steady-state sloping storms do not have to have mountainous terrain. All that's required is a change of wind with height. In the mid-latitudes, the jet stream can result in winds which increase with height over any kind of terrain. The greater the increase of wind with height, the larger and more severe the thunderstorms are likely to be, other things being equal. In the study

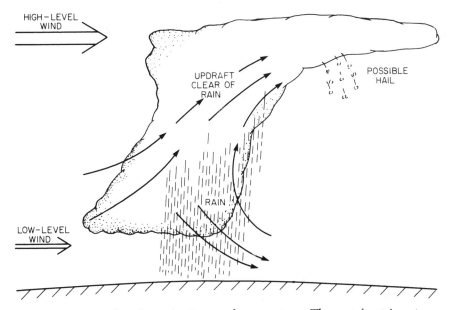

FIGURE 5-1 *A simple schematic of a steady-state storm. The gust front is not shown.*

of fifty-five general aviation accidents and eighteen air carrier accidents which we mentioned previously in talking about stability, the increase in wind with height was also found to be abnormally large.

A more dangerous variant of the steady-state storm which can occur anywhere that enough moisture and vertically changing wind exist is the back-sloping storm. The storm is an intense little low-pressure area, and air will flow into it from the sides. Dry air at midlevels, around 15,000 feet or so, may enter the storm from the rear and be invaded by rain falling out of the sloping updraft. Early in this process, the storm probably looks as shown in Figure 5-2. The falling rain evaporates in the dry air and cools the air in the process. The drag of the drops also gives this air a knock in the downward direction, resulting in a totally unstable situation. The air is cooler than the air around it as a result of extraction of heat by the evaporating rain, it has a downward shove to start it moving, it runs into still more rain to cool it even

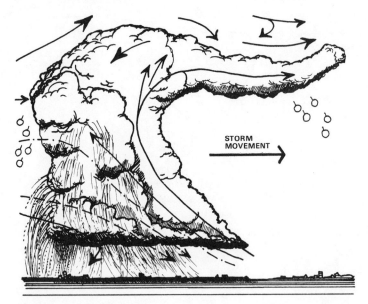

FIGURE 5-2 *A back-sloping storm; gust front not shown. (FAA.)*

more, it gets even more of a downward shove, and the worse it gets, the worse the situation becomes. This kind of storm produces hail just like the more simple orographic sloping storm which we found occurring in the mountains. It also produces a very dangerous gust front.

The air in the unstable downdraft, accelerating all the way through the cloud, roars out the bottom of the storm and moves out. The favored direction is ahead of the storm, since this is the direction of the prevailing winds which are moving the storm. Remember that this gust front is made up of cold air, cooled by evaporating drops as it came through the cloud. Thus, as the gust front races ahead of the storm, it plows its way under the surrounding air like a cold front. The lifted air is drawn toward the storm, and now stability gets into the act again. If the air at low levels ahead of the storm is moist and unstable, lifting ahead of the storm by the gust front triggers it into a building cloud which in turn enters the storm and feeds the updraft. This generates a stronger updraft, which generates more precipitation, which cools more air in the downdraft, which generates a stronger gust front, which lifts more air into the storm, which feeds the updraft, and so on.

Gust fronts have been observed more than 20 miles ahead of the generating storms. Peak gusts in excess of 70 knots have been observed 15 miles or more ahead of the generating storm, with sustained winds in excess of 40 knots following passage of the gust front. Once again, obviously, we don't have to penetrate these storms to encounter dangerous, even catastrophic, weather. We don't even have to fly to break an airplane. Leaving it untied on the ground in the path of a steady-state storm can do the job very nicely. If the wind doesn't get it, there's always the second chance that the hail will.

The steady-state storm is a genuine killer, and we penetrate one at the literal risk of our lives. It also requires more respect in circumnavigating than the air-mass storm. The question is, How do we know when we are looking at Mama Bear, not Baby? Some rules are as follows:

1. When in doubt, always treat the storm as a steady-state storm.

2. If you know that the lifted index in your area is −2 or less, unless you have very good evidence to the contrary, treat it as a steady-state storm.

3. The gust front advancing ahead of the steady-state storm and the blocking effect which such a storm produces in the upwind direction make these storms particularly inclined to form lines. This is true of all storms to some degree, but any line that forms rapidly and without plenty of space between storms should be treated as containing steady-state storms. Watch the storms that are in a loose line, like the ones we ran around in our air-mass flight, and see if the individual cells top out and dissipate.

4. Try to find out what the tops are. A call to Flight Service in the area may get you some radar tops, but be sure that the information is current and not read from a radar summary chart a couple of hours old. Two hours is ancient history around building thunderstorms. Listening in to the conversations on an ATC center frequency in the area may get you a report on the tops. As a ball park number, regard any storm with tops above 35,000 feet as a steady-state storm.

5. *Never* penetrate what you believe may be a steady-state storm. This also means not to penetrate a line of storms, in the clouds and unable to circumnavigate visually, without radar. Updraft and downdraft speeds may be in excess of 6000 feet per minute (I have personally measured an updraft of 5000 feet per minute in a cloud which had not yet reached the thunderstorm stage) and the wind shear between the drafts may be far in excess of what an airplane is designed for. This, of course, is in addition to the likelihood of encountering hail (see Figure 5-3*a* and *b*).

6. Exercise great care in taking off or landing at an airport in the path of a steady-state storm. The first gust'll getcha if you don't watch out. Fortunately, the leading edge of the gust front is often visible from the air. There may be a line of blowing sand or dust, trees will lean over and dance, and tall grass will show the edge clearly as it advances across a field. From the ground, though, even if you're looking for it, you may not be able to see the gust front coming until it's too late. The only good advice is not to take off if an advancing storm which you believe to be a steady-state storm is within 15 miles and you don't know where the gust front is.

As if Mama Bear wasn't bad enough, Papa Bear, the full-blown severe thunderstorm, is still waiting in the wings to be introduced. Let's let that bowl of porridge wait awhile, though, and take some time to understand the low-level wind shears which thunderstorms, particularly the steady-state storm, create.

Both wind speed and wind direction usually change with alti-

FIGURE 5-3a   *This buildup was one which I penetrated during a weather research project near St. Louis. It contained an updraft of more than 5000 feet per minute. (Courtesy of Dr. Bernice Ackerman.)*

FIGURE 5-3b   *Very shortly afterward it looked like this. (Courtesy of Dr. Bernice Ackerman.)*

tude. This is common and generally not hazardous. This is not what we mean by a wind shear, however. Since there are presently several definitions of wind shear around and since we don't want any confusion, we'll make our own. What we will mean by "wind shear" is a change in wind speed and/or direction which is sudden enough to cause an abrupt change in the indicated airspeed of an airplane. Let's look at an example (see Figure 5-4).

Suppose we have two layers of air, an upper layer moving from the left at 20 knots and a lower layer moving from the right at 20 knots, with a sharp boundary between them. The boundary we will call a shear line. Suppose you are watching from the ground, and an airplane is coming from the right, above the shear line but descending slowly, at an airspeed of 100 knots. What you see from the ground (we will ignore the difference between indicated and true airspeed, since it has no bearing on this illustration) is an airplane moving from right to left at a speed of 80 knots, but it has 100 knots of wind moving over its wings. As you watch from the ground, the airplane descends through the shear line. The wind change from a 20-knot headwind to a 20-knot tailwind is almost instantaneous. The airplane can't accelerate or decelerate instantly, though, so what you see from the ground, momentarily, is still an airplane moving from the right at a speed of 80 knots, but now with only 60 knots of air moving over the wings, since it has a 20-knot tailwind. Because the airplane can't accelerate or decelerate instantly, you, on the ground, haven't seen anything yet. You're about to, though, because the pilot has just seen the indicated airspeed drop abruptly from 100 knots to 60. The airplane was trimmed for steady flight with 100 knots of wind moving over its wings and control surfaces, and that's what it wants. Whether or not it actually stalls is probably of little consequence. In any case, it noses down abruptly in an effort to regain its trim speed. The pilot, any pilot, and any airplane, is now going to lose some altitude. How much depends on how much power is available, and how fast it is used, because what the pilot needs to do is reestablish steady flight by accelerat-

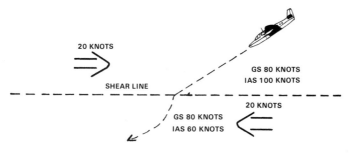

CASE ONE WIND SHEAR ENCOUNTER

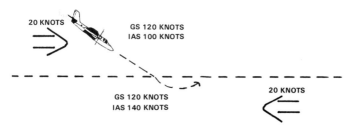

CASE TWO WIND SHEAR ENCOUNTER

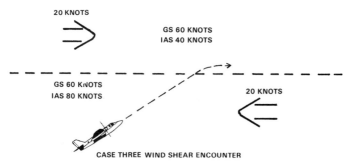

CASE THREE WIND SHEAR ENCOUNTER

FIGURE 5-4 *Idealized types of wind shear encounter.*

ing back to the trim airspeed, or at least to an airspeed which is sufficient to stop the altitude loss. If the airplane happened to be descending toward a runway, and if the shear line happened to be just a couple hundred feet off the ground, our pilot had best hope that the alligators in the swamp short of the runway are well fed. Let's call this kind of wind shear encounter Case One.

If our airplane was descending from left to right when it hit the shear line, the pilot would see a sudden increase in airspeed. You, on the ground, would see an airplane descending through the shear line at a speed of 120 knots, 100 knots of airspeed plus a 20-knot tailwind. The pilot is in an airplane moving over the ground at 120 knots, but with 100 knots of wind moving over the wings. When it goes through the shear line, you, on the ground, still see an airplane moving from the left at 120 knots (momentarily), but the pilot sees the indicated airspeed jump from 100 knots to 140 knots, the sum of the 120-knot groundspeed plus the 20-knot headwind. This is obviously not as dangerous as the previous case, since the airplane does not precipitate earthward, but as a result, the airplane may suddenly exceed landing-gear or flap-limit indicated airspeeds or overshoot a landing and wind up amid the stumps and alligators at the far end. Let's call this kind of wind shear encounter Case Two.

What we will call a Case Three encounter is as dangerous as Case One, perhaps even more so. Take the same wind picture we have been using, and suppose the airplane is climbing toward the shear line, from left to right, in the lower layer of air. Suppose it is climbing at its best rate of climb speed, for example, 80 knots. This is what the pilot sees on the airspeed indicator. Standing on the ground, you see an airplane climbing at a speed of 60 knots, with 80 knots of wind moving over the wings. Now the airplane penetrates the shear line, and what you see (momentarily) is still an airplane moving from left to right at 60 knots, but it now has a 20-knot tailwind, and therefore only 40 knots of wind moving over the wings. That's what the pilot is looking at, 40 knots. The airplane plummets earthward like a crowbar, but this time the pilot doesn't have any additional power to add, or at least not much. He was already using nearly all he had to climb. The wind shear will reverse itself when the airplane falls back through the shear line, but the airplane may be out of control or the sudden increase in airspeed may exceed limit speeds. If the shear line is close enough to the ground, it may not be possible to stop the descent in time.

For the sake of completeness, let's have a brief look at another possibility, Case Four, in which an airplane climbs through a shear line and encounters a sudden headwind. This is not usually dangerous, since it results in a sudden increase in airspeed and rate of climb. The airplane will pitch up in order to return to its trim speed, and as long as this is controlled and limit indicated airspeeds are not exceeded, there will usually be no harm done. However, a shear line is often violently turbulent and should be treated with great respect on that account regardless of the direction of penetration.

Now let's see what the thunderstorm, particularly the Mama Bear storm with the far-reaching gust front, can do to us. Look at the illustration and suppose we are landing at airport A (see Figure 5-5). The gust front has not yet arrived at this airport, but is approaching. The wind, however, is blowing toward the storm at airport A, feeding the updraft. This will usually be from a southerly direction, if the storm is approaching from the northwest quadrant. Let's assume that this is the case. Then, to make an approach to a runway oriented toward the south, we must fly to the north of the airport, more or less toward the storm. We are now set up for a Case One whammy, because we will very possibly enter an area above the shear line. We will start our descent toward the south in a headwind and descend through the shear line into the tailwind caused by the cold-air outflow from the storm. There may be some scud marking the shear line, or a roll cloud ahead of the storm rotating violently

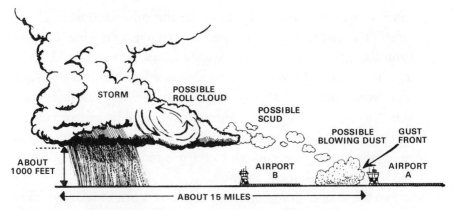

FIGURE 5-5 *A thunderstorm–gust-front wind shear situation.*

in the shear zone (see Figure 5-6). There also may be nothing visible. Winds in the southerly airflow at airport A and above the shear line may easily be as much as 30 knots, and winds below the shear line, in the cold outflow, as much as a steady 40 knots. This provides the potential for an airspeed loss of as much as 70 knots, accompanied by violent turbulence at the shear line. If we take off at airport A, we will take off away from the storm, and that's OK as long as the gust front doesn't pass while we're sitting there or on the roll. If it does, we can find ourselves attempting a takeoff in fierce turbulence with a 40-knot or so tailwind.

Suppose we are at airport B, and the gust front has already passed, as evidenced by a strong, cold wind blowing from the direction of the storm. In the example we are looking at, this would mean that we would take off (if we elect to try) to the north or northeast. We are now set up for a Case Three whammy, because we will climb through the shear line into the tailwind and again have the potential for an airspeed loss (accompanied by severe turbulence) of as much as 70 knots. That, obviously, would bring the airspeed to essentially zero in some light aircraft. The only good advice about taking off in the cold draft from the thunderstorm is, *don't do it.* If you do it anyway, turn away from the

FIGURE 5-6 *An approaching thunderstorm roll cloud.*

advancing storm as soon as possible. If you are heading away from the storm when you penetrate the shear line you will have converted the situation into the far better Case Four encounter. This, however, is a very risky thing to try to do, because you have no way of knowing how high off the ground the shear line is. If it's only a couple hundred feet, welcome to the stumps and alligators.

A Case Two wind shear encounter would occur in approaching airport B from the direction of airport A. If you know that the winds at airport A are blowing toward the storm, and that the winds at airport B are away from the storm, you know that the shear line is in there someplace. This time, you can expect fierce turbulence, strong moderate if not severe, and an increase in airspeed that may make landing impossible if encountered late in the approach.

Some pilots have played tag for years with thunderstorms in the vicinity of their departure or arrival airports and have gotten away with it. With experience and knowledge, we can be canny

FIGURE 5-7 *A gust front photographed near Miles City, Montana. (Courtesy of Mr. Lester M. Zinser.)*

enough to watch for the gust front from the air (see Figure 5-7), look for the scud or roll clouds which may indicate the shear zone, check winds from airports closer to the storm, listen to other aircraft on the frequency which may be flown by people more bold than we are and who are giving it a try, and otherwise make an intelligent decision on whether we are facing Baby Bear or Big Mama. All too frequently, however, such decisions are based on nothing but the desire to get going, or get there, and the outcome rests on the steady hands of dumb luck. Any airplane, and any pilot, can be brought down by low-level-thunderstorm wind shear, and the culprit is usually Mama Bear.

6 Will the Real Papa Bear Please Stand Up?

There is still one more kind of thunderstorm, the Big Daddy of the family, the full-grown severe thunderstorm. These thunderstorms generally travel in packs, complete with Mama Bears and Baby Bears, and woe betide the errant aeronautical Goldilocks who dips into the porridge. These are the things which spawn tornadoes.

These storms require large quantities of moisture and deep unstable layers to form. Typically, we picture an advancing low-pressure system in the midwest (see Figure 6-1). The low-level flow, "low level" meaning the lower 5000 feet or so of the atmosphere, is strong and steady and right off the Gulf of Mexico. These winds are frequently in the neighborhood of 40 to 50 knots in speed and are sometimes referred to as a "low-level jet stream." Tornadoes almost always occur in areas where the surface dew points are 53°F or higher.

A high-level jet stream from the southwest is usually found at 30,000 feet or so, crossing the low-level jet at an angle when viewed from above. This results in an environment in which the winds increase and turn with height, which is perfect for the development of steady-state storms. The updrafts will slope and the water brake will be off.

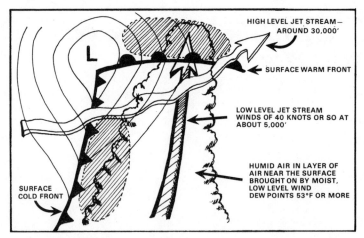

A Severe Thunderstorm Setup. Likely Severe Thunderstorm Locations shown by ▨▨▨.

FIGURE 6-1

Once the moisture and instability are there, and the wind field is favorable for steady-state storm development, all we need is lifting. This lifting is often provided, not by a cold front, as you might expect, but by lifting of the low-level jet stream air over a warm front. Squall lines and severe thunderstorms can and do occur with and ahead of cold fronts, but it is well worth making the point that tornadoes often occur along the surface of the warm front in the low-pressure weather system involved. This is directly contrary to the propaganda which is served up about warm fronts in most elementary weather courses for pilots, in which the warm front is generally presented in a stable situation with minimal convective activity. A warm front, like any other front, lifts air. If the air being lifted is a warm, wet, low-level jet stream and if the air is unstable enough and the wind changes with height enough, the result is not only thunderstorms, it may be Papa Bear, the severe storm.

Although isolated severe storms can and do occur, these storms are usually found in squall lines, and they start life as Mama Bears, steady-state thunderstorms with hail, gust fronts, ferocious updrafts and downdrafts, severe turbulence, and wind shears. As they develop into Papa Bear, they add yet another feature, the destructive tornado vortex.

One of the dangers of the tornado vortex is that it can be found in areas where you wouldn't expect it. Tornado vortices can exist aloft with no evidence at the surface, and can be found in areas with cloud tops as low as 12,000 feet. They do not have to extend vertically out of the storm but can be entirely in cloud and oriented almost horizontally.

As the steady-state storm develops, the updraft begins to rotate. Vortices of tornadic intensity can develop in, and extend beneath, the cumulonimbus cloud itself. This type of cloud can presumably exist as long as the conditions which created it (including the moisture supply) exist. The record for a single tornado path along the ground is over 7 hours. I won't even guess how long a vortex might exist aloft. New cells can develop on the right rear flank of the rotating updraft severe thunderstorm, and these have been found to contain tornado vortices. It is here, in the flanking line to the right rear of the severe storm, that the real trap exists (see Figure 6-2). These clouds are not Cb's, at least not initially. They may not paint on radar at all, and yet vortices have been seen extending from these clouds as far as 20 miles from the heavy precipitation of the thunderstorm. It is entirely possible to encounter a destructive vortex underneath this flanking line in clear air, miles from the storm, and while flying under a cloud which is not raining and which is clearly not a Cb.

The squall line which spawns the severe storm contains every kind of thunderstorm weather. The individual storms compete for the available moisture. Some may not become thunderstorms at all. Many will, and Baby, Mama, and Papa Bear will most likely all be found in various places—updrafts in excess of 6000 feet per minute, downdrafts of that, or more. I have heard of one case (in which the airplane survived) of a 12,000 foot per minute downdraft—large hail, gusts in excess of 70 knots near the

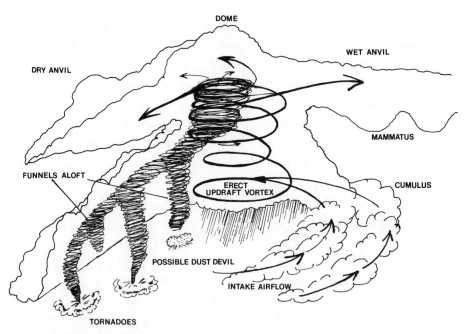

FIGURE 6-2 *The Bates model of the severe thunderstorm.*

ground, severe and possible extreme turbulence, and tornado vortices.

What is a tornado vortex? Everyone has seen pictures of a tornado tearing up the turf in the midwest someplace. The width of the tube may be anywhere from a couple of hundred feet to a mile. Tubes may be found in or under the Cb, and in or under the flanking cloud which is connected to, but not yet part of, the storm in the upwind quadrant. The tube may be oriented anywhere from vertically to nearly horizontally and will be visible only if it is carrying dust or debris of some kind. The speed of the winds may be over 200 knots in the vortex, and grazing or penetrating one can create any kind of wind shear. A head-on encounter can literally lift the wings of an airplane right out of the fuselage. There are cases on record, noted particularly by the late Dr. Fred Bates, in which airplanes apparently disintegrated in clear air in the vicinity of severe storms, probably due to a head-on vortex encounter. A tail-on encounter can result in an airplane suddenly finding itself going backwards relative to the wind. There is a case on record, again from Dr. Bates, in which a jet transport at about 19,000 feet in a flanking line found that its indicated airspeed had abruptly become zero, whereupon it did the only thing any sensible airplane with no airspeed ever does: It put its nose down and got some. The airplane survived due to some beautifully sharp piloting, but the recovery pulled one engine right out of the wing.

It is probably not necessary to go on with stories of what could, and has, happened. Note, however, that these vortices very seldom contain water. Unless they are continuously picking it up from the surface, the rotation throws it all out. Consequently, since weather radar "sees" water drops, they do not paint on radar. It is possible to penetrate a squall line containing severe storms, to be well away from the nearest storm echo, and to hit a tube.

Fortunately, this most destructive of storms is perhaps the most easily anticipated. The National Weather Service has a method of forecasting these storms. It involves the high- and low-level jet streams, low-level moisture, stability, and lifting and is one of the single most reliable methods of forecasting. The National Weather Service issues teletype severe weather watches whenever necessary. The coverage area is normally a box, 60 nautical miles or so each side of a line between the endpoints of the suspect area. These boxes are then drawn on the radar summary charts which are received by Flight Service and National Weather Service stations. The example shown contains one severe thunderstorm box, noted on the chart as WS167, and three tornado watch boxes, WT168–170 (see Figure 6-3).

These forecasts verify about 40 percent of the time; that is, the forecast weather occurs in the box during the forecast time period. Of those which don't precisely verify, many are near misses, with the weather occurring a little outside the box or outside the time period. Also, for these forecasts to be verified,

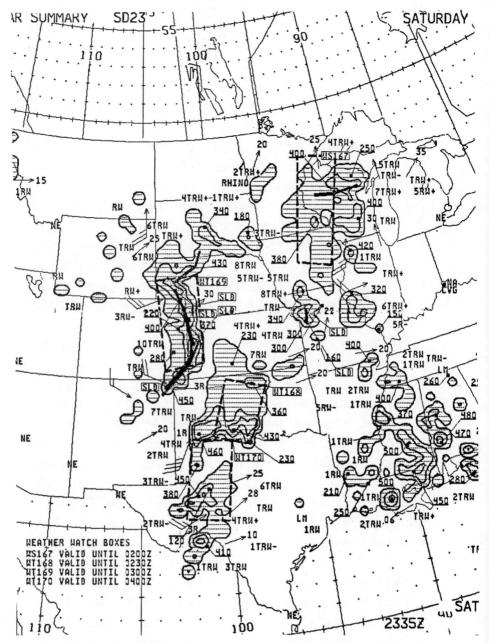

FIGURE 6-3 *A radar summary chart showing severe-thunderstorm-watch and tornado-watch areas.*

the weather generally has to affect the surface. The fact that this may not happen in an individual case doesn't mean that the weather doesn't exist aloft. Therefore, avoidance of Papa Bear, the severe thunderstorm, generally means just this: When a severe weather watch is in effect and storms have actually begun to form, stay out of the box. Development in these areas is so rapid (do a little arithmetic with a 6000 foot per minute updraft and see what you get) and penetration is so risky even in clear air, once storms are present, that to go out of the way or wait a few hours is an insignificantly small price to pay for the assured safety it will afford.

Suppose that a severe thunderstorm occurs without being forecast or that word of the forecast has not reached you. All the tips we discussed in the chapter on Mama Bear, the steady-state storm, can still be used to determine that this is no ordinary storm. If you are caught in the area and have to circumnavigate what you suspect may be a severe storm under VFR, avoid it by at least 20 miles. Don't run under a cloud base upwind of the storm, and even more emphatically don't if the cloud is seen to have a common base with the storm. For the radar-equipped instrument pilot, a severe storm may give a couple of clues, as follows:

1. Watch for right-moving storms. A large thunderstorm echo which turns to its right, instead of moving in the direction you would expect from the wind flow, is probably a severe thunderstorm.

2. Hook-shaped echoes often indicate storms containing tornado vortices.

The right-moving characteristic of some severe storms might also be observed with a Stormscope, if you are so equipped. Any storm with an abnormally high frequency of lightning flashes, which of course would be observed with the Stormscope,

FIGURE 6-4 *Approaching the flank of the Wichita Falls tornado-producing storm. (Courtesy of Bill Wood and Sperry Flight Systems.)*

FIGURE 6-5 *The Wichita Falls squall line. The tornado storm is dead ahead at a range of 20 miles. (Courtesy of Bill Wood and Sperry Flight Systems.)*

should be treated as a steady-state or severe storm.

Once again, give such storms a 20-mile berth. Upwind the particular hazard is the tornado vortex; downwind it is hail and gust front turbulence. Never fly through, or under, the upwind flanking line. Dr. Bates once calculated that penetrating a flanking line in the upwind direction could result in odds of about 1 in 6 of hitting a tornado vortex. That, as he was fond of reminding people, is the same odds as Russian roulette.

On April 9, 1979, a Cessna 421 operated by RCA with a photographic setup on their Primus 300 color radar got some spectacular shots of the severe storm that sent a tornado through Wichita Falls, Texas (see Figures 6-4 to 6-6). The pictures show beautifully the development of the tornado-producing cells in the flanking line upwind (southwest, in this case) of the mother storm. The airplane approached to within about 10 miles of the storm in clear air and was struck by dirt and small stones picked

FIGURE 6-6 *The Witchita Falls storm. Note the weak radar returns from the flanking lines. North is roughly at the top of the scope. (Courtesy of Bill Wood and Sperry Flight Systems.)*

up by the storm. The tops of the cell directly off the nose appeared to be no higher than 15,000 feet—all in all, as nearly perfect a photograph of Papa Bear as we may ever expect to get.

7 The Downburst

On June 24, 1975, an Eastern Air Lines 727 crashed during an instrument approach in thunderstorm conditions at JFK, killing 113 people. The subsequent investigation, which included detailed analysis of data from the airplane's flight data recorder, directed a great deal of attention to the wind shear conditions which were encountered. The term "downburst" was coined by Dr. T. Theodore Fujita of the University of Chicago, and he defined it to mean a localized, intense downdraft with vertical currents exceeding a downward speed of 730 feet per minute at 300 feet above the surface. The downburst results in an "outburst," where the vertical air current hits the surface and spreads out violently. The fastest spreading flow, again from Dr. Fujita, is seen in the direction of the cell motion.

Dr. Fujita's analysis of the available meteorological data identified a thunderstorm which produced a "spearhead echo," moving much faster than the parent echo, as the culprit in the accident. The top of this thunderstorm was apparently in excess of 40,000 feet.

Attention was soon directed to some other similar accidents which had occurred previously. In addition, there have been at least three subsequent similar accidents and incidents which have been analyzed. Simulators have been programmed with flight data recorder information derived from these aircraft to examine whether the conditions were in fact beyond the performance capabilities of the airplanes involved. The conclusions have been that the chances of recovery from such a low-altitude downburst are marginal at best, even during a simulation when the pilot knows what's coming.

As a direct result of this, wind shear detection equipment which compares the wind at the center of an airport with the winds measured at other locations around the field was devised. A Jeppesen Briefing Bulletin, which is provided to subscribers of their chart service, listed twenty-four major airports having such equipment in operation as of January 25, 1980.

Recently, however, researchers at the University of Dayton Research Institute have come up with some things which seem to me to be of good practical use in downburst avoidance. They concluded that in nearly every case of a thunderstorm-related air carrier accident, the aircraft penetrated a heavy rain cell during final approach. My own reading of NTSB reports of several of these accidents bears this out. These reports also, in several cases, detail problems that other aircraft had in heavy rain just prior to the accidents. The University of Dayton people even put forth the idea that the effects of the rain itself on the airplane may be of the same order of magnitude as the wind shear effects, and they correctly point out that the analyses and simulations done on these accidents made the assumption that the only problem was air currents and would therefore have overlooked any problems due to the rain. They believe that the Eastern 727 could easily have encountered a rainfall rate as heavy as 12 inches per hour in addition to the wind shear and that the rain could have degraded the performance of the airplane in several ways, making recovery impossible.

I don't propose to try to arbitrate any dispute which there may be about the relative effects of the rain and downburst or other wind shear on the airplanes involved in these accidents. However, it comes as no surprise to me that heavy rain was encountered immediately prior to practically all of these accidents. It only serves to reinforce the conclusion that taking off or landing through rain from a thunderstorm is potential suicide. The only safe policy is, *don't do it.*

8 *Lightning*

Is lightning hazardous to airplanes? You bet it is. There have been many, many lightning strikes on airplanes that have done no more damage to the airplane than a small hole somewhere in the skin and no damage to the crew that can't be repaired in a laundromat. Once in a while, though, something like you see in the picture happens. In this case, about the last 6 feet of wing came wide open at the first line of rivets aft of the leading edge (see Figure 8-1). Once in a while, also, an airplane goes down. Lightning strikes on aircraft are far more common than most people imagine. Air carrier airplanes, for which statistics are available, are struck on the average of once about every 3000 hours. Let's take a look, from the standpoint of present knowledge, at what a lightning strike is, what sort of damage it can do to an airplane, and what a pilot can do to minimize the chance of being hit.

Although I'm not at all sure that everyone involved in lightning research would agree, I'm going to go slightly out on a limb and say that there appear to be at least two kinds of lightning strikes to aircraft, which I will call (1) "The Random Zap" and (2) "The Other Kind." In at least The Random Zap case, the mechanism is basically as follows. A very high voltage, low-current leader emanates from a highly charged cloud, creating

FIGURE 8-1 *The leading edge of a wing near the tip after being blown wide open by lightning.*

a relatively short ionized conductive path through the air. The leader advances in steps on the order of 100 to 150 feet at a time, at a speed less than 1 percent of the speed of light. It may go in various directions and may fork a number of times before the end of one of the forks reaches opposite charge on the ground or in cloud. Once this circuit is completed, a very high current electric discharge, called the "return stroke," takes place at about one-third the speed of light along the ionized path and may be followed by several lesser strokes. These return strokes are what the eye sees. How much current? As much as two hundred thousand amps is possible! Most, however, are less than fifty thousand amps, with an average on the order of probably twenty thousand. Most of the current in a given return stroke will be over with in something like one-thousandth of a second. If an airplane happens by within a fairly short distance of the leader as it is forming (probably within a few hundred to a few thousand feet, although an exact distance would be nearly impossible to specify in any given case), the leader may attach to the airplane and then go on its way toward opposite charge. If the branch going through the airplane is the one which connects, all the return-stroke current will flow through the airplane. The lightning will not terminate on the airplane, which doesn't begin to have the capacitance to handle the charge.

So what can happen? The worst possible thing is a fuel explosion. Liquid fuel will not explode, but if a combustible mixture happens to exist in the tank airspace or in a fuel vent, a spark would make it possible. While no fuel can be considered completely safe, the safest common fuel in this regard is Jet A, which will not naturally have a flammable mixture in the tanks in the altitude and temperature region in which a lightning strike is likely. One hundred–octane avgas is less safe. The most likely flammable regions occur at high altitudes and cold temperatures, 30,000 feet and −40°C for ball park numbers. Jet B and JP-4 are much less safe. Even with the "safe" fuels, however, agitation by turbulence can increase the possibility of a combustible mixture existing. Mixing of different types of fuels in the tanks can also increase this possibility. An airplane well-designed for lightning-strike tolerance is ordinarily designed to prevent sparks in these areas regardless of the actual likelihood of a combustible mixture existing.

The second worst thing that can happen is turbine engine flameout. This is not uncommon with small turbine engines, particularly those aft mounted along the fuselage. There is at least one case of a double flameout of an airplane with aft-mounted engines on record. These cases of engines quitting are thought to be due to disturbance of the airflow into the engine as the return stroke sweeps past the inlet. No damage is usually found and an in-flight relight of the engine is generally normal. However, pilots of small turbine airplanes should be alert to this possibility when operating in the vicinity of lightning and should be ready with the airstart checklist in case it's required.

Beyond this, a number of different types of structural, avionics, and electrical damage are possible. The leader, when it first approaches an airplane, will draw electric discharges from metal parts of the structure or electric wiring at the ship's extremities, i.e., nose, wingtips, propeller tips, pitot masts, and so forth. If the discharge to which the leader attaches happens to come from something inside a nonconducting fairing, such as a fiberglass wingtip, the current in the return stroke will probably vaporize most of the nonconducting material. The buildup and decay of the return stroke or strokes occurs very rapidly, causing magnetic forces capable of doing a good bit of metal bending and inducing voltages in the ships wiring to run around and burn things up. Turning unneeded electric or avionic equipment off in anticipation of a possible lightning strike may well fail to protect it from a strike.

Well, what can we do then? In the case of The Random Zap, not much except avoid areas of thunderstorms. Anywhere in or out of cloud in the vicinity of, and especially between, active cells, a Random Zap is possible. If you feel it coming, close one eye. Then if the flash blinds you temporarily, you won't be any worse off than Wiley Post, who did all right for quite a while.

What I have chosen to call "The Other Kind" of strike, however, may be quite a bit more avoidable. The Other Kind can perhaps do everything The Random Zap can do, although it seems to be generally milder. However, it usually occurs in cloud, within a temperature band of -5 to $+5$ °C, in precipitation, and with some degree of turbulence. Over 80 percent of lightning strikes for which there is data seem to be The Other Kind. The conclusion, of course, is obvious. At the very least, if you are going to operate around thunderstorms, you will do your lightning-avoidance chances no harm (and perhaps a lot of good) by:

1. Avoiding the freezing level by at least $\pm$ 5°C
2. Staying out of cloud as much as possible

One final word on the subject. NASA has published an excellent book called *Lightning Protection of Aircraft*, by Fisher and Plumer. It is NASA Reference Publication 1008 and is available in paperback from the Government Printing Office. It contains 550 pages on lightning protection of everything from light single-engine airplanes to the Space Shuttle, and while some of it is fairly technical, it is copiously illustrated and most of it can be readily understood by pilots and mechanics. It's aimed primarily at engineers and designers, but I recommend it highly to anyone who operates airplanes around thunderstorms and wants to know how well protected they are against lightning hazards.

9 ***Thunderstorm Weather Briefing***

We have said a few things in passing about weather briefing for flight on thunderstorm days. I don't care to repeat them, but there are some things which haven't been said, so let's take a look at how to obtain a weather briefing that'll give us the confidence that comes with knowing what we're doing when we walk out to the airplane.

If possible, go in person to the Flight Service station, or whatever other weather shack you may use, on days when the weather is questionable. A picture is indeed worth a thousand words, and you will both save time and increase your understanding of the weather on this day and all other days if you go see for yourself.

First, and this is common to all weather briefings and has been said so many times it's almost trite, take a look at the big picture. In the case of thunderstorms, first take a look at the stability chart, if you can find it. It is seldom used or asked for in weather briefings, and that's a shame. Many times you will see immediately that the chance of thunderstorms is small, and you can concentrate on other things. Other times it will alert you at once to a potentially explosive situation. In any case, it's a lot of quick and painless thunderstorm information. This chart comes off the facsimile machine as one panel of a four-part moisture chart, and it is prepared twice daily using 00Z and 12Z data. It does not come off the fax, unfortunately, until almost 7 hours later. Even 17Z, however, varies from noon to 3 P.M. over the continental United States, so it's still in time to be of use. Even the previous chart is worth looking at, because the stability parameters frequently do not change that rapidly. Perhaps it will become possible to get this chart a little sooner now that automation of Flight Service stations is in progress and charts can be called up on the tube. I hope so.

Now go to the surface map. What we particularly look for here are moisture and lifting. The moist areas are the areas of high dew point. We pay particular attention to areas where the dew point is 53°F or higher, because that's the magic tornado number, and to all areas where the dew points are more than 40°F or so and lifting might occur. As to lifting, we look for several things. We look for large cloudless high-pressure areas where the ground will heat up in the afternoon to get thermals started. We look for fronts and note what they have been doing for the last several hours if the previous charts are available. We look for squall lines, which often form along instability lines away from fronts (see Figure 9-1). In the example shown, there is a squall line (shown by a dash and two dots, repeated along the length of the line) out ahead of a cold front. This is a common place to find a squall line, but certainly not the only place. A squall line will be shown on the surface map by this dash-dotted symbol whenever it is not coincident with a front. The dew point is the lower left-hand number in each group of numbers around a station, and we see in this example that the dew points are in the high 50s and low 60s, well in advance of the frontal system and squall line.

If it's available, a look at the low-level winds-aloft chart would be handy at this point. We will look at the winds above the surface up to 5000 feet and see if there is a general pattern of wind from large bodies of water, particularly from the Gulf. On a day such as shown on the example surface map, it would not be unusual to find strong southerly flow from the Gulf to the Great

Lakes, with still more water from the Lakes being added to be lifted over the warm front at the top of the map.

The radar summary chart is next. We look for areas of organized activity, and we look at tops. We also look at the time the chart was prepared and take the information with a grain of salt. Remember that 2 hours is a long, long time when you are talking about things that can grow at over 6000 feet per minute. We also look at the radar summary chart for severe weather boxes, to alert us to areas where a severe weather watch was current at the time the chart was prepared. We look at the intensity level of the storms, which is determined from the strength of the radar returns. Level 5, in particular, is a very intense storm.

The charts are not the latest weather, so we now go to the hourly sequence reports and check stations along our route. Hav-

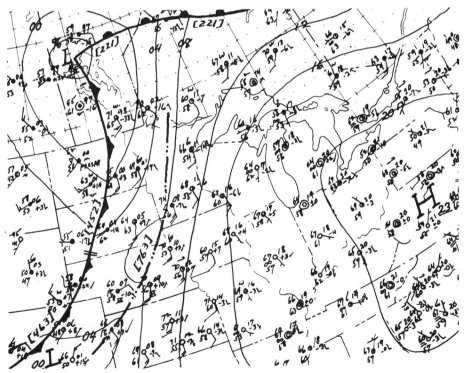

FIGURE 9-1 *A surface weather map showing a prefrontal squall line. (FAA.)*

ing checked the weather charts, we know the direction we can expect weather which might affect our route to come from, and we also check existing weather out there. We check dew points and existing cloud layers which might be cumulus or altocumulus. We check the remarks at the end of the reports for mention of towering cumulus clouds or Cb's which might be within sight of any of the stations. We check pilot reports for tops of the haze and for any sighting of thunderstorms.

We must be sure to ask for, and check, convective SIGMETs. For several years now, at 35 minutes past each hour, the Kansas City Severe Storms Forecast Center has been issuing a convective SIGMET bulletin on the teletype. If there are no thunderstorms to report, the SIGMET is issued anyway and merely says none. This is an excellent service. (Unfortunately, it could have been an even better one. In April 1978, the FAA issued what was called a "prototype hazardous weather plotting chart" with radials and distance circles drawn on it with respect to six VORTACs. The chart covered the entire United States, and the convective SIGMETs had the position of the storms with respect to the plotted VORTACs. The purpose of the chart was to allow easy in-flight plotting of the storms by pilots receiving the SIG-MET, and I thought it was great. However, it soon fell by the wayside.)

Anyway, whether there is any present activity on the convective SIGMET or not, we look for severe weather watches. We may know from the box on the radar summary chart that a severe weather watch has been issued, but the absence of a box on the chart does not mean that one couldn't have been issued since, so we check. We then look at the prog charts for the time we are concerned with and determine what the surface map is expected to look like at that time. We check these charts for expected areas of thunderstorm activity. Finally, we zero in on our destination and possible alternates by checking terminal forecasts.

This last step particularly requires bringing to bear the knowledge we have gained from our preliminary checking. The forecasts were always made some time ago, or they wouldn't be laying there on the table. Obviously, they were made by the forecaster from knowledge of weather which existed at some still earlier time. We will look at what the weather is forecasted to be right now and see how the forecast is panning out so far. If we are to be scared off by such things as "chance of" or "possible" thunderstorm in the forecasts, we might as well resign ourselves to walking from April to November. If thunderstorms are definitely indicated by the forecasts, we will look for the moisture, lifting, and instability which will cause the storms to develop. Things may have changed since the forecasts were made.

We also consider the possibility that the forecasts do not predict thunderstorms which will happen. If the moisture and instability are there, and a little lift comes along, the atmosphere won't ask permission from the forecast. Remember that, due to the scale of the data-collection system, thunderstorms are often like little fish in a big net. The strands of the forecast network may have missed something that is now bumping against the strands of the current data-collection net, and the fish that we could've found and didn't will be no less voracious as a result of our oversight.

I take no pleasure in exercises in 20/20 hindsight. I once served as a meteorological consultant to the NTSB on a weather accident in which a good guy who I liked a lot died, and that's grim. However, knowledge of weather accidents is valuable for its educational value, so let's consider one. If you will take another look at the radar summary chart in Chapter 6, you will see one severe thunderstorm box, WS167, and three tornado-watch boxes. Within an hour after this chart was made, an airplane broke up in flight in WS167, having apparently been subjected to more than six G's (see Figure 9-2). The stability chart which would have been available for weather briefing prior to the flight shows

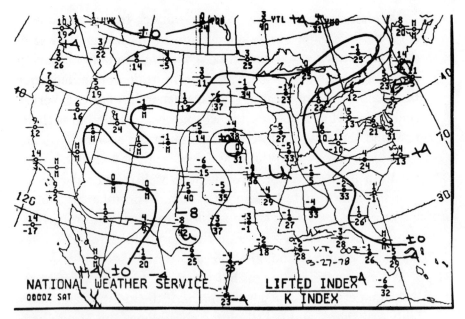

FIGURE 9-2 *The stability chart as it came off the facsimile machine before the accident. To the old list of a pilot's most worthless things might be added weather information you don't see.*

a lifted index of -1 in the area of the accident, rapidly increasing to -5 a little farther south. The K index is very high and positive throughout the whole area. As might be expected, the testimony from the Flight Service folks who briefed the pilot (which is contained in the NTSB report) makes no mention of any use of this chart. In addition, the pilot was apparently never told of the existence of the severe thunderstorm watch, even though he contacted Flight Watch for weather information less than 10 minutes before the accident. The watch was almost 3 hours old at that time, and he was right in the middle of the storm. Briefed on the existence of thunderstorms he was, including a current sequence report of a heavy thunderstorm with surface gusts to 50 knots about 40 miles from his position and moving his way. The preflight briefings had also included the probability of thunderstorms along the route, so the pilot certainly can't be said to have been unaware of the storms. All the same, when you've missed the tremendous instability and the severe thunderstorm watch, it seems to me that you've thrown the baby out with the bath water. The only last word on checking weather for thunderstorms is as follows. If there is a severe-weather watch current, believe it! They're good. Find out where the area is, and stay out of there. Nobody, but nobody, needs to mess with Papa Bear.

10 *Nocturnal Thunderstorms*

It may be well to say a few words about thunderstorms which occur at night, because the subject is sometimes overlooked or treated as obscure in weather books commonly available to pilots. There is nothing basically mysterious about nocturnal thunderstorms. All three kinds of storms which we have talked about here can, and do, occur at night. I don't have any data to support this, but I would guess from my experience that air-mass storms occurring at night are more often unforecast than daylight storms and that the consequent element of surprise lends a sense of the unknown to these things.

They occur for the same reasons that any other thunderstorm occurs, namely, the right combination of moisture, instability, temperature, and lifting. There are far fewer thunderstorms at night than during the day because of lack of surface heating by the sun. This removes the thermal lift which is probably the most common trigger for air-mass storms, and if the ground cools off enough it cools the layer of air near it resulting in a warm over cold, stable, layer. However, it's easy enough to imagine situations which can kick off air-mass thunderstorms at night. Let's imagine.

If the air is moist enough and unstable enough above what may be a shallow, stable layer near the ground, a high enough hill can always do it. If enough low-level wind springs up to move the air up the hill until a cloud forms and if the lifting continues until the air being lifted becomes warmer than the air around it, Bingo!

Suppose air which is sufficiently moist, and is unstable except possibly in a layer near the cool ground, is moved by the wind over a relatively warm lake. Not only will the lower layer warm up and tend toward a cold over warm, unstable situation, it will become even more moist. A very small amount of lifting caused by rising terrain on the lee shore may then be enough to set off the action.

On the other hand, absence of wind from any other sources can result in formation of a sea breeze front at night, in which relatively cool air from the land moves offshore and plows under the moist air over the water like a miniature cold front. Once again, possible fireworks.

In the right wind field, and with enough moisture and instability, any of these storms can become a steady-state storm. Steady-state and severe storms can also be set off at night by more conventional lifting mechanisms, i.e., cold fronts plowing under warm, moist air or warm, moist air riding up over warm or stationary fronts. Steady-state storms which form during the day can also last far into, or even through, the night.

A cloud which formed during daylight may cool off at the top after sunset, again resulting in a cold over warm, unstable, situation. A small upper-air trough may form or move across an area where air is sufficiently moist and unstable for thunderstorms. These are hard to locate on the large scale of meteorological upper-air charts, but they are not hard to understand. They do one or both of the following:

1. They act like a small vacuum sweeper in the upper air and suck up air from below, thereby providing lift.

2. They bring in cooler air aloft, and there we go again—cold over warm.

Thunderstorms, even surprise thunderstorms, are fortunately often more easily avoidable at night than daytime storms because they broadcast their location, and frequently give a good indication of their intensity, by easily visible lightning. Once again, get on top of any haze, if possible. Don't panic at the first flash of lightning, because it may be visible for hundreds of miles at night, but do start immediately to use all the means at your disposal to locate the storm. If the lightning is frequent or covers a broad area, suspect steady-state storms and possible squall lines and act accordingly. Don't hesitate to land if you don't like the situation. The last time I did that the local sheriff came out to investigate, and we spent the rest of the night drinking coffee and telling yarns until I could be off in the morning. It sure beat flying those storms.

11 *Thunderstorm Detection Equipment*

It is beyond the scope of this book to try to give a course in the use and interpretation of thunderstorm detection equipment, and it might not be wise to do so anyway. This is an area in which a little knowledge can truly be a dangerous thing. The manufacturers of these goodies provide some generally excellent manuals on how to use them, and there are also several good courses and seminars available. I can't recommend too strongly that any pilot who uses this equipment obtain this sort of detailed information on what it's good for (and not good for). However, it would probably be remiss not to address some of the questions about the merits of such equipment which commonly arise, so let's do it.

There are two types of equipment available. The first is radar, which is available in many different flavors from several manufacturers. Radar has, of course, been around for years. There is at least one British patent on radar for ship detection and navigation dating back to 1904. Radar has the advantage that there is a wealth of experience in scope interpretation, and anyone who learns how to operate and interpret it can do an excellent job of weather avoidance. The newer digital sets can also be made to do additional useful tricks like display checklists and other information, and there are now radar installations available for high-performance single-engine airplanes. While the mathematics of radar design can become mind-boggling, the principles involved are very simple. The word radar stands for *RA*dio *D*etection *A* nd *R* anging, and the device is based on the principle that some of the energy in radio waves is reflected by the objects they strike. Therefore, if you want to locate some object and learn something about it, you throw radio waves of a wavelength appropriate to the size of the object at it and you measure the strength of the returned energy and the amount of time required to make the round trip. In the case of weather radar, the objects are drops of water. A wavelength which works well for this purpose is 10 centimeters. Energy transmitted at this wavelength will penetrate storms at long ranges and still return enough energy to be used for a scope display. The only problem is that this wavelength requires an antenna over 10 feet wide. For obvious reasons, use of this wavelength is pretty well limited to ground-based radar.

Airborne weather-radar equipment generally operates at either the so-called C-band (5-centimeter) or X-band (3-centimeter) wavelength. Most are X band, which are smaller and lighter than C band. Radio energy at these wavelengths is reflected well from raindrops. The larger the drops and the more of them there are, i.e., the harder it's raining, the more energy will be reflected. Radar manufacturers have taken advantage of this fact by programming the display scopes to show the higher energy returns as brighter contours or in different colors and by showing areas of rainfall rate known to be associated with dangerous turbulence either in red (on color scopes) or as black holes, sometimes flashing. The elevation angle of the antennas is pilot-controllable, so that some idea of the slope of a storm and of its top can be had. The rainfall gradient can also be easily seen and used for avoidance of turbulence, since areas where the rainfall rate increases rapidly over a short distance (high-rainfall gradient) are also known to be associated with turbulence. All these things make radar a very useful severe-weather-avoidance tool (see Figures 11-1*a* and *b* and 11-2*a* and *b*).

The drawbacks of airborne weather radar, in my opinion, are the following:

FIGURE 11-1a and b *Airborne radar being used as intended, for storm avoidance. In this case, the storm being painted by the radar is also plainly visible with the old eyeballs, as shown.*

FIGURE 11-1b

FIGURE 11-2a and b *Another radar paint perfectly suitable for storm avoidance. In this case, however, the picture out the windshield shows that visual storm avoidance would be impossible.*

FIGURE 11-2b

1. *Attenuation.* The capability of X-band radar, in particular, is extremely sensitive to degradation due to precipitation which may be falling between the antenna and storms farther out and also due to radome icing. According to the NTSB reports, failure to appreciate this problem probably contributed heavily to the crashes of a Southern Airways DC-9 in 1977 and an Air Wisconsin Swearingen Metro in 1980. In both cases, ingestion of very large amounts of water and/or hail caused loss of power on both engines of each airplane. In the case of the Metro, the NTSB's analysis of the situation based on available recorded information from ground-based radar in the area indicated that the X-band radar on the airplane would not have begun to show contours until the flight was within 1 mile of the storm because of rain attenuation and that the total detection capability of the radar would have been reduced to less than 9 miles after the heavier rain was encountered. The long and short of it is that airborne weather radar is very good at looking at storms from clear air, and thereby allowing the pilot to avoid them, but it can really snooker you if you try to penetrate precipitation with it (see Figure 11-3).

2. *Ground Clutter.* The earth, and objects on it, also reflect energy back to the antenna. It can be very tough to pick the weather out of the hills, particularly if you want to do it from the ground before takeoff.

3. *Cost.* Radar sets ain't cheap. However, sitting on the ground because of weather ain't cheap anymore either, and you'll certainly do less of that.

4. *A Certain Amount of Radiation Danger.* Turning the set on while sitting on the ground can expose nearby people to hazardous radiation. Care should be exercised when looking at weather before takeoff, and radar sets should be turned off after landing to avoid injuring anyone.

The second type of thunderstorm detection equipment presently available is the Ryan Stormscope, now manufactured by

FIGURE 11-3 *A scope picture not to trust. The airplane is flying in rain, and the open area shown behind the bright return just off the nose may not be there. It may be the result of attenuation.*

3M Company. The Stormscope, being much newer than radar, is perhaps of more interest in terms of comments on its usefulness. It is not my intention to be a drum beater for the Stormscope, but I am happy to testify to its utility. Although its exact workings seem to be kept pretty close to the vest, published information indicates that the Stormscope generally works by sensing the intensity of radio-frequency discharges in the atmosphere through what amounts to an ADF antenna and comparing these discharges with a theoretical lightning discharge model contained in its computer. The computer decides what is a lightning discharge and what isn't and displays those which pass its tests as dots on a CRT screen. As the screen fills with dots, it updates itself by throwing out the oldest ones and replacing them with new ones. A pretty good idea of the relative severity of a storm can be had by the rate of lightning discharges occurring in it.

The Stormscope has several advantages over radar: It costs much less. It is not affected by ground clutter, and it is really nice when sitting on an airport in mountainous terrain to be able to inspect the area around you for activity. It can be installed in virtually any airplane. It also has a longer range than airborne radar, and the range is unaffected by altitude or rain attenuation. It radiates no energy, so there is no radiation hazard. Its drawbacks, seem to me, to be as follows:

1. What it senses is the electromagnetic field from a lightning discharge, and it computes the range to the strike. How it does this I do not know, but it has to be making some assumption about the strength of the discharge, which as we have seen can vary all over the place. Practically speaking, however, I have not found this to be a serious limitation. There are many, many lightning discharges in any given storm, and the picture painted by the Stormscope seems to be adequately accurate for storm avoidance on the whole.

2. There are well-known correlations between types of radar return and turbulence. There are no known correlations at present, or at least none that I would bet on, for the Stormscope. I would not use it for any close penetration of areas of activity unless I could stay out of the clouds. (That doesn't mean that such correlations do not exist, however. A good bit of research is being done on that subject as of this writing.)

So there you are. You pay your money and you take your choice. Both radar and the Stormscope are very useful pieces of equipment for staying out of thunderstorms, and either will enable you to make better use of an airplane by giving you reliable real-time information on weather you want to avoid. It just depends on you to use them wisely and, like the rest of your airplane (and yourself), not to depend on them to do something they can't.

12 *Hows and Whys of Icing*

After all these years of flying in ice, studying it, and designing and building equipment to cope with it, icing problems it seems are still with us. It is probably the weather hazard second only to thunderstorms in the opinion of most pilots who spend a lot of time at relatively low altitudes. It even managed to contribute heavily to bringing down a 727 a few years ago, because of an anti-icing system which was not turned on. We'll take a look at icing from several points of view. We'll discuss how it forms on an airplane, and where. We'll see what icing clouds are, where they exist, and why. We'll discuss the present sorry state of the art in icing forecasts. Then we'll look at how to anticipate icing clouds from data available on the ground and how to recognize and avoid them in flight. Finally, we'll discuss the effects of ice on an airplane, how to cope with the effects, and what the legalities of flight in icing conditions are.

The accretion of ice on the components of an airframe is caused by and the form of the ice is controlled by several meteorological and aerodynamic factors. These factors include cloud liquid-water content, temperature, cloud-droplet size, and the size and speed of the ice-collecting object.

The most important meterological factor in the generation of ice is cloud liquid-water content. The amount of ice which will accumulate on an airframe component is directly proportional to the amount of water contained in the cloud in the form of liquid droplets. Water contained in the cloud in the form of vapor, snow, or ice crystals will generally not stick to the airplane and contributes little or nothing to the overall ice buildup. The second most important meteorological factor, which affects both the likelihood of icing and the form the ice will take, is the temperature. The vast majority of icing encounters of any significance occur with outside air temperatures between 0 and $-10°C$. Below $-20°C$ icing is a very rare event. This is true largely because the likelihood of finding high liquid-water content in a cloud decreases rapidly as the temperature decreases. This is a statistical conclusion, however, and should not be taken as gospel in any specific case. It is possible to get plenty of ice at colder temperatures, as we will see in a couple of exceptional cases which will be discussed later. It is also possible to get ice at ambient temperatures above $0°C$ on parts of the airplane where there is a pressure drop, such as engine-induction areas. The warmer the temperature at which ice is encountered, the worse the shape of the ice is likely to be, as will shortly be seen.

The size of the cloud droplets have a significant effect on both how much ice will form and what shape it will take. The effect on total accumulation occurs because of what is called "collection efficiency." As the airplane moves through the air, it disturbs the pressure field around it. This results in a pressure wave moving out ahead of each component at the speed of sound. Nearly everyone who flies has seen the effect of this in some sort of wind tunnel picture. The airflow, warned by the pressure disturbance, parts and moves smoothly around the disturbing object. The cloud droplets also try to make the turn with the airflow. To oversimplify a bit, the small ones make it and the big ones don't. Therefore, the larger the drops, the more ice you get. The collection efficiency is the ratio of the amount of ice an object *does* collect to the amount it *would* collect if it caught every liquid droplet in its path.

If you spend a minute thinking about this, you will quickly arrive at the conclusion that the size of the object collecting the ice, as well as the drop size, has to be a factor in how much ice will form. This is absolutely true. However, the effect of size is actually the opposite of what might be expected at first glance. The larger the airframe component, and particularly the larger its leading-edge radius, the less ice it will collect. This is true because the larger objects give the air ahead more warning that they are coming and allow the air to get out of the way with more gentle turns which carry more water along, causing more water to miss the component. The effect of speed is just the opposite. The faster the component is moving through the air, the less warning the air has that it is going to have to scramble out of the way. This, of course results in more of the water failing to make the sharper turn and smacking into the airframe. (Do not, however, draw the conclusion that slowing down is the thing to do when ice is encountered. Actually, it would be the thing to do if it could be done without increasing the effective frontal area of the airplane, but it can't. The increase in angle of attack will make matters worse, not better.)

By now, it will be readily apparent that all the above factors combine in a complicated way to influence aircraft icing. It should also be apparent why ice collection has to be considered in terms of components rather than just in terms of the aircraft as a whole. We can make the following generalizations of the effects of the various factors in combination:

1. *Amount of Ice.* If suitable liquid-water content and temperature conditions are present, the small components moving at high speed in the presence of large drops will collect the most ice. The large components moving through relatively small drops at slow speed will collect the least ice, and possibly none. Typical collection efficiencies for a light twin in an "average" icing cloud are about 45 percent for the wing leading edge, 85 percent for the empennage leading edges, and 95 per-

cent or more for small components such as pitot tubes and OAT probes.

2. *Shape of Ice.*   The shape of the ice which forms depends on all the meteorological parameters we've discussed. The aircraft speed is also a factor, since it influences the temperature of the ice-collecting object. Warm temperatures, large drops, and high liquid-water content are the worst combination. When the ice accumulation begins, large and relatively warm drops run back and freeze aft of the point on the leading edge where they strike. More drops then freeze on the ridges formed by this process. The final result is a buildup with a more or less horn-shaped cross section which is aerodynamically awful (see Figure 12-1). Small drops at cold temperatures tend to freeze where they strike, resulting in a spearhead type of formation. This may affect lift nearly as badly as the horn-shaped ice, but it is much less damaging to drag.

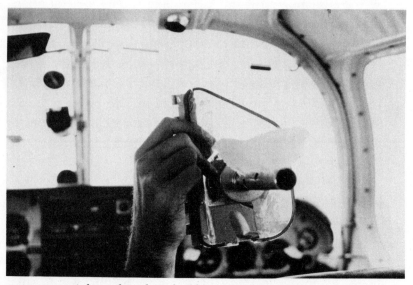

FIGURE 12-1 *A horn-shaped ice buildup accumulated on a temperature probe which was installed in the pilot's openable window during icing flight tests. The window was brought inside the airplane to allow an ice sample to be preserved.*

3. *Type of Ice.*   The two shapes of ice just described are what are commonly thought of as clear ice and rime ice, respectively. However, there is no sharp transition between these two types of icing. Since higher speed raises the temperature of the leading edges and increases the rate at which water strikes, it is entirely possible to penetrate a cloud at low speed and collect a rime formation and then to penetrate it at higher speed and collect an intermediate or clear-ice shape. There is also, obviously, a point at which aerodynamic heating is sufficient to prevent ice formation, at least on leading edges.

4. *Location of Ice.*   Some of the remarks made about the amount of ice an object will collect lead to the conclusion that icing will be different on different parts of an airframe. Small parts will collect ice first. It is possible to ice a pitot tube without icing leading edges. It is also possible to ice the leading edge of the empennage with little or no icing on wing-leading edges.

To identify and avoid icing conditions requires a basic understanding of how various types of icing clouds are formed. It is then possible to learn to look for areas of probable icing, both on weather charts and through the windshield. The worst of it can generally be avoided if you know where to look, what to look for, and what to do.

If you don't want ice, the first and most important thing to do is stay out of clouds with high liquid-water content. The primary cause of high liquid-water content is lifting of moist air (the Fundamentals again), which results in cooling and the condensation of large amounts of water vapor into liquid drops. Once liquid-water drops exist in the cloud, their primary enemy is the creation of ice crystals in the cloud, called "glaciation." Water goes from vapor to ice in a cloud with great difficulty, but it goes from liquid to ice very easily. Once glaciation starts, the cloud engages in a battle between upward air motion creating more liquid drops and glaciation converting the liquid water to nonliquid crystals of ice. In the absence of energetic lifting of air, a cloud will glaciate in just a few minutes. The colder the temperature, the more likely it is that glaciation will begin and the more quickly the cloud is likely to become dry. Clouds have been penetrated for flight test purposes at temperatures as warm as −3°C, resulting in heavy ice buildups, and then repenetrated less than an hour later and found to contain nothing but snow (see Figure 12-2*a* and *b*). The following can be said about various types of clouds:

1. Cumulus clouds with relatively warm temperatures at the cloud base are likely to be very wet in their building stages. As a ball park number, developing cumulus clouds can produce very heavy ice buildups if the temperature at the cloud base is −12°C or warmer. The warmer the cloud base, the worse the possibilities.

2. Stratus-type clouds are far less likely to produce severe icing. Obviously, there is much less vertical motion going on, or they wouldn't be stratus. The vertical extent of the icing layer will be much less in stratus than in cumulus. Icing layers more than 3000 feet thick are rare. The highest liquid-water contents and largest drops are often found near the tops of stratus decks. Fast ice buildups are unlikely with a cloud-base temperature of less than −3°C. The horizontal extent of an icing encounter in stratus is much greater than in cumulus, but still generally less than 30 miles. *Exception!* None of this necessarily applies to lake-effect stratus, formed usually in early winter by cold northwesterly winds blowing across the unfrozen Great Lakes and then over land. There is so much moisture available that these clouds can sometimes stay ahead of glaciation and remain sopping wet well inland as long as the conditions which created them continue to exist.

3. Stratocumulus clouds in relatively thick layers may be wetter than stratus, but not so wet as cumulus. This, naturally, is because the lifting is between the two pure types in intensity. Most of the ice is likely to be found in "lumps" in the cloud deck, where the lifting is the greatest, and near the top of the deck,

FIGURE 12-2a *A glaciated cloud. Notice the fuzziness at the edges.*

FIGURE 12-2b *Another glaciated cloud.*

where the lifting will have squeezed out the most liquid water. Lake-effect clouds may take this form, as well as the pure stratus form, and the same warnings apply. Icing layers are still seldom more than 3000 feet thick.

4. A type of cloud which I don't know to have been studied by itself for icing conditions is the wave cloud. Experience has shown that they are very wet. The reason is simply that the lifting in the upward moving parts of the waves is sufficiently energetic to squeeze out a lot of water, and these conditions can be very long lasting (see Figure 12-3). The direction of flight through a wave-cloud deck will obviously have a great influence on how much icing is encountered. Icing will be intermittent and of short duration if the flight proceeds perpendicular to the waves but can be disastrous if the airplane flies along the wave in the area of lift.

5. Cloud-droplet size has an effect on icing, for the reasons mentioned previously. Droplets tend to be larger in cumulus than

FIGURE 12-3 *A wave-cloud day, photographed near Philipsburg, Pennsylvania, which produced plenty of structural ice.*

in stratus. In addition, the cloud-droplet size is likely to be considerably larger in clouds that form over open water or in clean air than in clouds which form in dusty or polluted air. Consequently, clouds that originate over or draw their moisture directly from bodies of water are likely not only to be wetter but to contain larger drops.

13 *Icing Weather Briefing*

The object of the icing game is to avoid areas, and altitudes, where the clouds are likely to be icing clouds and to spend the least possible amount of time in such clouds when they must be penetrated. This is a chore which has two parts. First, it involves study of the weather situation and judicious preflight planning. The second step is to recognize potentially hazardous weather in flight and to know what to do to stay out of it, or to get out of it. When we come to preflight planning, we almost immediately come to the subject of icing forecasts. The kindest way to put it is that they are not very good.

The good news is that NASA has become interested in icing. Again, actually. They sort of thought they (NACA at that time, of course) had solved that problem back in the fifties. They pretty much did, in fact, as far as transport category airplanes are concerned. What has really spurred the current interest is the problem of helicopter icing, which turns out to be a whole different bag. It also turns out that the icing environment of helicopters is very similar to that of many general aviation airplanes and that the problems in these areas are nowhere near being solved. NASA held two meetings relating to icing during 1978. The first was a workshop in meteorological and environmental inputs to aviation, which was held at the University of

Tennessee Space Institute in March. It examined, among other things, severe storms, turbulence, visibility, lightning, and icing. More or less as a result of this workshop, an unexpectedly grand soiree totally devoted to icing was held at the NASA Lewis Research Center in Cleveland, Ohio, 4 months later. The attendance was beyond the sponsors' wildest dreams, with representatives coming from Canada, Britain, France, Sweden, West Germany, Norway, and The Netherlands, as well as from all over the United States. They came from government(s), aircraft manufacturers, universities, research and consulting firms, various associations, and the military, well over one hundred of them. They gave the whole state of the art a good going over, splitting off into groups which considered meteorological research, icing forecasting, ice-protection-systems development, civil and military operations, and icing research facilities. The various groups met with each other and ground their own particular axes, much figurative (but no literal) blood was shed, and a grand time was had by all.

Now, of course, the bad news. The consensus at both affairs was that icing forecasting isn't much good. The icing committee at the March conference reported that icing forecasts (primarily as judged by the Army—there was no comparable civil data available) are felt to be accurate only about 50 percent of the time. The weather services committee at the same meeting reported, "There is an urgent need to improve the capability to forecast icing conditions . . . ," and recommended National Weather Service action on the subject.*

At the July conference, a representative of the NTSB made a presentation showing that statistics indicated, in most cases, that icing was forecast in the area in which an icing accident occurred. Any conclusion which might have been drawn from this presentation that the forecasting was satisfactory, however, was almost immediately assailed by representatives of both general

---

*The report is NASA CP 2057, FAA-RD-78-99, available through the National Technical Information Service, Springfield, Virginia 22161.

aviation (including yours truly) and the U.S. Army. There was no statistic available on the other side of the coin, i.e., how many times icing had been forecast and none of consequence had been encountered. The icing forecast committee at this conference, which, like the March committee, included folks from NASA, FAA, general aviation, universities, and the military, also concluded that improvements in icing forecasting are required.

It is just as appropriate for those of us who fly to understand the limitations of weather forecasts as it is to understand the limitations of our equipment. We have already mentioned some of the physical idosyncracies of icing clouds which make them difficult to forecast. However, it turns out that there are two more problems with icing forecasts which make them virtually meaningless. The problems are definition and verification.

The definitions of icing conditions used by the National Weather Service and FAA are in a state of absolute chaos. The following National Weather Service definitions come from an airframe icing-reporting table (*Airman's Information Manual: Basic Flight Information and ATC Procedures*):

TRACE: Ice becomes perceptible. Rate of accumulation slightly greater than rate of sublimation. It is not hazardous even though deicing–anti-icing equipment is not used unless encountered for an extended period of time—over 1 hour.

LIGHT: The rate of accumulation may create a problem if flight is prolonged in this environment—over 1 hour. Occasional use of deicing–anti-icing equipment removes/prevents accumulation. It does not present a problem if the deicing–anti-icing equipment is used.

MODERATE: The rate of accumulation is such that even short encounters become potentially hazardous and use of deicing–anti-icing equipment or diversion is necessary.

SEVERE: The rate of accumulation is such that deicing–anti-icing equipment fails to reduce or control the hazard. Immediate diversion is necessary.

These definitions relate the intensity of the icing condition only to the ability of the ice protection equipment on the airplane to handle the problem. Two otherwise identical airplanes, one with and one without an ice protection package (which is not unrealistic, by the way—there are a number of airplanes on the market for which ice protection is partially or totally optional), could fly through the same cloud in formation and then report light icing and severe icing, respectively. Both would be perfectly right according to the table.

These definitions are useless for forecasting and were never intended for that purpose. In order to forecast the icing, you would first have to forecast the airplane and the equipment on it. Yet, incredibly, they are the only official definitions the National Weather Service uses. A point which was made over and over again at both 1978 NASA conferences was the necessity to define icing intensities in terms of cloud characteristics, not airplane characteristics. Unfortunately, a subsequent NASA report entitled "Light Transport and General Aviation Aircraft Icing Research Requirements," dated March 20, 1981, indicated that no improvement had been made in the situation.

We can also add some other confusion which is found in the Federal Aviation Regulations. FAR Part 135 and Subpart D of Part 91 allow aircraft with ice protection equipment certified to transport category (FAR 25) criteria (which as of this writing is still the only way to certify ice protection equipment regardless of whether the airplane it's on is transport category or not) to fly into severe icing conditions. This contradicts the National Weather Service definition, by which severe icing conditions are impenetrable. FAR 135 also allows some airplanes which have certain specified items of deicing and anti-icing equipment installed, even though not FAR 25–certified, to fly into known or forecast moderate icing conditions. Since the definition of moderate icing conditions varies from airplane to airplane and is not suitable for forecasting in the first place, how are we supposed to give a forecast of moderate icing any confidence at all? For the record, FAR 25, the certification rule, and FAR 121, which deals with airlines operating transport category airplanes certified to FAR 25, do not even mention intensities of icing conditions.

The second problem, the problem of forecast verification, arises in part from the definition mess. The National Weather Service definitions allow, nay, require, that two differently equipped airplanes report two different icing intensities in the same cloud. In addition, such data as have been taken using precise instrumentation have generally been taken by airplanes either seeking icing and loitering in it when found, or deliberately avoiding it as much as possible. The result of all this is that icing is about the only thing for which forecasts are made where there is absolutely no way of judging forecast accuracy. To quote Laurel and Hardy, who seem as appropriate to quote on this subject as anyone else at the moment, "A fine mess this is."

Even the simpler question of whether any icing condition regardless of intensity will exist has not been adequately answered as yet. The U.S. Army helicopter fleet in Germany was reported at the March 1978 NASA meeting to be grounded about 30 percent of the time during the winter because of forecasts of icing conditions, and yet it has had numerous icing encounters when no icing forecast existed. Manufacturers interested in certification testing have searched for ice using National Weather Service forecasts and gotten none whatever at times. On the other hand, I remember a flight from Baltimore to Hickory, North Carolina, in a Cherokee not too long ago for which no icing was forecast because the temperature was supposed to be $+2°C$ or higher all along the route. It turned out to be $-2°C$ instead, and we started loading up. Considering the terrain and the stratus-type clouds which indicated that descending would not bring us to warmer air, I asked for and got an immediate 2000-foot climb. That put us out of the liquid area of the clouds, but we had picked up about one-half inch of ice on the wings and more on the tail in a hurry, and we kept it all the way to the final approach at Hickory.

The service is also nothing sparkling. I had the experience of standing in the National Weather Service office at Seattle-Tacoma International Airport faced with an icing forecast which I did not believe during a flight-test program and found that there was no real-time data available which I could use to form my own opinion. All the data was downtown at the forecast office, which was too busy to provide any detailed information over the phone since that is not its function. The forecasts are made downtown and the airport crew blindly reads them. They didn't have the information necessary to evaluate them. In most cases of this sort, they would not be permitted to comment on the forecasts anyway. This led to the absurd situation of a flight crew calling the Penn State University weather lab on the phone from the SEA-TAC National Weather Service office to get icing information for Seattle. National Weather Service headquarters was later very cooperative about arranging access to the downtown office, and everyone there was also very cooperative in the ongoing flight-test operation, but that doesn't help a transient pilot trying to get some information at the airport one iota.

Problems like this were also brought up at the two NASA conferences. The official answer seems to be the coming Automation of Field Operations and Services (AFOS) System, which will get rid of teletype and fax machine data and send weather info to CRT displays via computer hookups. A number of responsible people from the National Weather Service and FAA who were there defended and promoted this concept of increased automation as the answer to acknowledged weather briefing inadequacies. I don't question for a minute their commitment to providing improved weather service or the extent of their efforts in that direction. However, I wish to state my opinion that *it's not working.* Instead, in the case of icing and other types of weather as well, it is moving the pilot just one step farther from the raw data, the sequence reports from all over the area for the last several hours, the last couple of surface maps, the low-level upper-air charts, and the last few radar summary charts, which a few years ago a pilot could study until the pilot felt confident that he or she understood the situation. We are increasingly spoon-fed forecasts backed up with less and less exposure to the data which they come from, even when we have good reason to doubt them. In the case of icing forecasts, for the reasons mentioned earlier and which came to light at the NASA workshops, we have no good reason *not* to doubt them. I, for one, am in no way convinced that automation won't just make it that much harder, instead of easier, to look at the data for ourselves.

The forecast committee at the July 1978 NASA conference concluded that improvements in the forecasts will be forthcoming only when the icing definitions are changed to involve quantitative and forecastable cloud conditions instead of airplanes, and when instruments are available to measure and report these conditions in terms which can be compared with the forecasts to determine their accuracy. We can't wait that long to fly. The forecasts are virtually useless in avoiding icing. The usual case is that the forecast will be there whether icing is or not. The opposite case, icing with no forecast, has also been known to occur. In fact, I got a good certification flight-test icing encounter on one occasion on a day when the National Weather Service had forecast no icing in the area. Well, then, what is better?

What is better is, unfortunately, not the simplest thing to do. It requires going, in person, to a Flight Service station or National Weather Service office and then hoping that they have the information you need. It can be done over the phone, but it is usually like pulling teeth. Most weather briefings consist of a description of the surface map and forecasts which are read as if infallible, and little more. That is nowhere near enough. If you want to assess an icing situation, you must do four things. Sure enough, the Four Fundamentals again. First, find out where the moisture is. Second, in the areas where there is moisture, evaluate the temperatures. Third, if there is an area along your route with suspicious moisture at the right temperatures, find out what is going on that might lift the air. Fourth, get some idea of the stability of the air, i.e., how well it will cooperate with being lifted.

The place to start is not with the surface map but with the upper air. Look at the 850- and 700-millibar charts and, if you will be going that high, at the 500-millibar chart. These are altitudes of about 5000, 10,000, and 18,000 feet MSL, respectively. There is a little circle at each station with data plotted around it. If the temperature–dew point spread at a given station is 5°C or less, the circle will be shaded in. If there are shaded-in stations along or near your route, look more closely. Check the temperatures. If they are between $+1$ and about -15°C, and particularly if they are between 0 and -10°C, there may be trouble. If the temperatures are in the danger range, check the temperature–dew point spreads more carefully. If you find an area where the temperatures are right and the spreads are 2°C or less, treat that area as extremely suspect. Compare the same stations on the different altitude charts and see if they are moist at all levels. The temperature or the moisture, or both, may be safe at one level but not at the other. Look at the temperature difference between the two lower levels. If the 700-millibar level is more than 7°C colder than the 850-millibar level, you have a cold over warm case in which the air in that area may be, or become, unstable.

These charts are, unfortunately, history. Sometimes they are ancient history. The data is taken at 0000Z and 1200Z, and is a couple of hours old when the charts are first accessible. Therefore, you will not only want to know where the moisture was when the chart data was taken but where it was coming from and where it was going. There are two things to look for: If it is still available, look at the chart made 12 hours previously for the same level. Where was the moist area then? Did the area become larger or smaller between the two data times? Did it become wetter or dryer? Colder or warmer? What you want is the trend, less serious or more serious.

Whether or not the previous charts are available, look at the winds on the current charts. Are the winds which are moving the moisture coming from warmer areas, or colder? Wetter, or dryer? Will the temperature and moisture probably get better, or worse? Make a particularly good check of the low-level winds. This can be done with the 850-millibar chart and the surface map, if necessary, but the low-level-winds aloft charts present the information more clearly if they are available. What you particularly want to spot is a wind field at low levels coming over land from a large body of water. Considering the worst of the ice traps in the United States, this means easterlies or northeasterlies in the Atlantic coastal area, northwesterlies over the eastern Great Lakes, and westerlies in the Pacific northwest. These winds bring moisture, and there is terrain to lift the air. Southerly winds from the Gulf of Mexico also carry lots of water and will provide the raw material for plenty of ice when the temperature is cold enough. Even if the current chart does not show moisture, the next one will if one of these wind conditions exists.

All this may sound like the object is to make each pilot a forecaster. That is, of course, not possible and not the intent. The intent is to give each pilot who takes the trouble the necessary tools to be a good observer, by pointing out where to look and what for. By following the preceding steps with the upper-air charts, more or less one potato, two potato, you can find answers to the following questions: Where were the areas of suspicious moisture (temperature–dew point spread of 2°C or less) and temperature (0 to −10°C, in particular) recently (at the chart time)? Which way were they traveling? Were they getting larger or smaller, more moist or less? Can they be circumnavigated? Can they be topped, or "bottomed"? Is there likely to be a lake-effect trap? Which way out?

With a little practice in reading the data, the check of the low-level upper-air charts described above can be done very quickly, 5 minutes or so. Then, having had a look at where there is, or possibly will be, enough moisture to be troublesome at the right temperatures, we have to look for mechanisms which will lift the air. These come in two varieties, meteorological and orographic. The latest surface map and prog charts are the place to look for the primary meteorological lifting mechanisms, which are low-pressure systems and fronts. The more energetic the lifting is, the wetter the clouds which are created will be. In this regard, it is especially necessary to understand and appreciate the three-dimensional nature of weather. The airplane obviously operates in all three dimensions, going up and down as well as thisaway and thataway. Having checked the upper air in at least the lower 10,000 feet for moisture and temperature conditions conducive to icing and having flagged areas of possible instability by looking at the temperature difference between the 850- and 700-millibar levels, we can see where the lifting caused by the features shown on the surface map is likely to produce the wettest clouds. If the areas in the neighborhood of a front are wet and/or if the winds bringing in the air ahead of a cold front or behind a warm front are coming from a body of water, watch it. In either case, the place to expect the most ice is from some distance ahead of the surface position of the front to some distance behind it. It is impossible to specify a distance in general, because it will depend upon the amount of moisture available, its distribution in both horizontal and vertical space, how fast the front is moving (i.e., how energetic the lifting is), and what the stability and temperature of the air is. This is why, if you want to intelligently exercise the pilot's prerogative of route planning to avoid icing, or bothersome weather in general, you absolutely must have some idea of what the low-level upper air looks like.

Orographic lifting, i.e., lifting by airflow over rising terrain, can cause some of the most hazardous icing conditions. Other things being equal, a front won't do anything a mountain range won't do. All that either of them really does is lift air. Furthermore, if you are going to get a load of icing where there is no icing forecast, there is a good chance that it will be orographic. We spot this case primarily by looking for a low-level wind field blowing over a body of water and then uphill. Two classic cases are westerly winds from the Pacific Ocean moving over the Olympics and the Cascades and northwesterly winds from Lake Erie and Lake Ontario over the inland mountain ranges. These are probably the two worst ice traps in the United States. However, orographic icing conditions can obviously also exist over many other mountainous areas when the moisture and temperature are right. One common and easily overlooked source of moisture is lots of water laying on the ground from the passage of some sort of storm system a few hours previously.

The wave cloud is a special case of the development of icing conditions due to lifting air by terrain, and probably the worst of the orographic types. The typical wave-cloud situation will occur when a wind field about perpendicular to a mountain range develops after the passage of a cold front through the area. The moisture may come from the ground, dropped there by the cold frontal system. Also, in the northeastern and west coastal mountain ranges, the low-level wind field after a cold frontal passage often comes directly from a large body of water. The low-level

FIGURE 13-1 *This photo was taken during a mountain wave research project downwind of the Blue Ridge. We were supposed to be doing this in clear air. These clouds weren't supposed to be there. (Courtesy of Dr. Ron Smith.)*

air behind the front is typically unstable because of the cold air temperature over relatively warm land. This results in the available water going rapidly into the air. The winds running perpendicular, more or less, to the ridges combine with the temperature structure typical of the situation to create sopping-wet wave clouds over and downwind of the mountains (see Figure 13-1).

The orographic situations are typically at their worst from late morning to late afternoon, because the sun's heating helps lift the moisture. If wave conditions exist, some degree of avoidance is possible by planning trips to avoid these times. Depending on where the trip is going, studying the 850- and 700-millibar charts and the wind fields may enable you to simply avoid the area where conditions are particularly favorable for wave lifting. If your airplane is capable of it, an altitude 5000 feet or more above the ridges will usually top the wet orographic clouds.

But enough theory. Let's assume that you've done the best you can with preflight planning using the available information and are on your way. Now the problem becomes a very practical one, which we'll consider next.

14 *In-flight Icing Avoidance*

Avoiding ice, or getting out of it if you have been unsuccessful at avoiding it, really boils down to just two things. Either get to an area where the temperature is warm enough to melt the ice or get out of the liquid parts of the clouds.

The temperature part looks elementary at first glance, and it is really not very complicated. There are a couple of traps worth mentioning, however. If you are getting ice and an altitude below you has been reported by another aircraft to be ice-free because of warmer temperatures, the obvious solution is to go there. Before you do, however, try to find out some details about the PIREP. If it was made by a large turbine airplane tooling along at 250 knots and you are flying a Twin Zorch with some ice on it at 125 knots, you may get ice that they didn't get because of their aerodynamic heating. An indicated airspeed of 250 knots will protect leading edges, at least, down to true outside air temperatures of −6°C or so, which is to say, a lot of the time. See if you can find out what the real OAT is down there. Otherwise, you take a calculated risk by going down, and you give up precious altitude. If you leave an altitude where you are picking up a fairly benign ice shape and go to a warmer one where, instead of getting rid of it, you begin to collect horn-shaped ice, you have done yourself no favor.

You can infer something about temperatures above and below you by the kind of clouds you see. If they are very flat stratus types, and the air is smooth, the temperature will probably change very little if you go up or down in them. If there are energetic cumulus buildups, the temperature will warm up 2°C or more per 1000 feet if you descend. Above all, do not blindly use the standard lapse rate. The standard lapse rate is science fiction. It is an engineering standard used to calibrate airspeed indicators and altimeters and has nothing to do with weather on any given day.

There is a lot that can be done to identify and stay out of highly liquid clouds. Let's start with the assumption that you are out of cloud for the moment and looking at them ahead, around, or below. The first rule is, *stay out of building cumulus.* Avoid them at icing temperatures like you would avoid a thunderstorm. If you have to penetrate them, do it under the most favorable conditions, which are at temperatures either warm enough to prevent ice formation or as cold as possible. All cold-temperature bets are off in very energetic Cu's or Cb's, however. They have been known to be wet down to temperatures of −40°C, and possibly colder, as a result of the large amounts of water condensed by the lifting.

The second rule is, *stay out of wave clouds.* These are easily identified from above, and they will distribute copious amounts of ice all over your aerodyne at the right temperatures. If you have to penetrate them, holler and scream at ATC to let you stay over or under them until you can be cleared to go all the way through, then do it. Do not go parallel to the ridges down through one of the areas of rising air, i.e., upward-moving wave, or you will wind up icier than a double martini.

The third rule, related to the first two, is obvious. *If you must go through, penetrate any suspected or known icing cloud by the shortest possible route, either horizontally or vertically, or both.*

The fourth rule is, *read what the clouds are telling you.* Spend a minute looking at cumulus clouds and see if they are building upward. Those clouds, or parts of clouds, are the most likely to be wet. See if the edges of the clouds are sharp and well-defined, or diffuse and fuzzy. The first characteristic indicates liquid water, the second indicates already frozen ice crystals. If the sun is in the clear above you and you are on top, you may be able to tell directly if the clouds below are liquid. Look at the cloud tops in the down-sun direction, where the airplane shadow would normally be. If you see colored rings around, or instead of, your shadow, the cloud is liquid (see Figure 14-1). If you don't see this, look down on the clouds in the other direction, toward the sun. If you see a brilliant spot of light on the clouds, not colored, you are looking at a reflection from ice crystals. The top of the cloud deck, at least, is not liquid (see Figure 14-2).

Now, suppose you don't have the option of peering at the clouds from safely outside them. Suppose you are in them. The previous rules still apply, of course; they are just more difficult to follow. Two more rules, in this case, are, *be ready* and *spot the ice early.* Have your available anti-icing equipment on when

FIGURE 14-1 *These colored rings, called a "glory," indicate that these clouds are liquid.*

you penetrate possible icing conditions, whether you actually expect icing or not. Look at some small object exposed to the airstream for the earliest possible indication of ice formation. An OAT probe is good. So is a thin antenna mast. The empennage leading edge is better than the wing leading edge. Remember, the smaller the exposed object, the higher its collection efficiency, and the sooner ice will collect. If there is no small object at all visible in the airstream, consider installing one if you are going to fly in icing conditions. Don't put it where ice which sheds from it will damage something, however (see Figure 14-3).

If you are in cloud, you obviously can't avoid the wetter types visually. If you have radar and you paint cells at an altitude in the icing-temperature range, by all means avoid them. However, even very wet (for icing purposes) cumulus clouds will often not paint because the drops are too small. You will have to detect this type of cloud by the seat of your pants. The bumps will tip you off, and if there are updrafts, be especially watchful for ice. A single icing cumulus cell is generally quite small, a couple of

FIGURE 14-2 *The bright reflection on the cloud tops, called a "subsun," indicates glaciated clouds.*

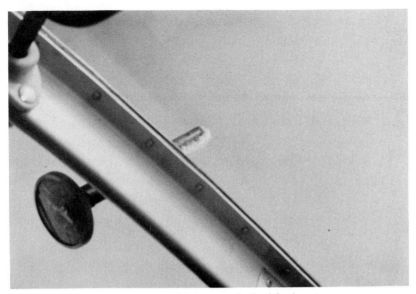

FIGURE 14-3 *A first indication of icing can be had by keeping an eye on some small exposed object, like this OAT probe.*

miles or so across, so your immediate best course of action if you start collecting ice is usually to bore through. Without good ice protection equipment, however, you can't loiter in a wet cumulus cloud very long or go through a bunch of them without picking up more ice than anyone wants. If you find yourself caught in this situation and you can get on top of the general cloud cover and then pick your way around the Cu, climb quickly while you can. If you are sure that there are ice-melting temperatures at a safe altitude below you, go on down there. Remember, though, that sooner or later, voluntarily or otherwise, you are going down there anyway, and there is no point to hurrying it if you are not sure it will be warm enough. If you continue encountering cumulus cells depositing ice on your vehicle and you know or can find out the directional orientation of the cause of the lifting which is setting them off (e.g., a front or a ridge line), proceed away from the situation by the most direct route.

Wave clouds are more difficult to identify if you are in them.

Strong ones are detectable, however, if you know what to look for. In the first place, if you are flying over or downwind of a ridge line when the wind flow is more or less perpendicular to the ridges, suspect them. Some day when favorable conditions for waves exist without enough moisture to create bothersome clouds, look for them. When you find them, play with them a little. The best way is to fly toward the ridges a couple of thousand feet above them and do what is necessary to hold altitude. If you have an autopilot with an altitude-hold feature, turn it on. Fly for 20 miles or so and watch. A moderate case of wave motion will result in slow airspeed variations, sometimes very smooth, of ± 15 knots or more as you maintain altitude. Variations of ± 10 knots are common even in relatively weak waves. This will also be true in clouds, of course, if there are clouds, and may serve as your tip-off on what is going on. If you are flying more or less parallel to the ridge line and your airspeed to hold altitude seems slow, you may be in a downward-moving area of a wave. If your airspeed seems high and conditions are otherwise favorable, suspect that you are in an upward-moving wave. This, of course, is the likeliest area to be wet as a result of the lifting. If you proceed along an icing wave in the lifting area of the cloud, your airspeed won't be high long. The ice will take care of that. Again, an important part of the solution is to catch the ice buildup and analyze the cause, early. The ways out are to get the airplane headed in a direction not parallel to the ridges and go up or down, as appropriate. If you can climb 4000 feet or so, up will generally do it. That is usually preferable to down unless you know the temperatures down there will melt the ice off. Remember, you will be going down anyway, unless you are in orbit.

If you decide that you are just in everyday stratus, an altitude change of 2000 feet in either direction will probably do it and 4000 feet will almost certainly do it. Stratus not involved with lake effect or waves is not very wet through any great depth. As far as up versus down goes, stratus does tend to get wetter as you approach the top of the cloud deck. Otherwise, the remarks made about climbing and descending in other cloud types apply.

An important characteristic of icing is its extreme variability in both space and time. There are good reasons for this. An airplane going right through the updraft part of a wet wave cloud will pick up a terrific load, and an airplane 10 miles away at the same altitude but in a downdraft area may get nothing. If the air is relatively cold, one airplane may get a load of ice in freshly formed clouds and another airplane on the scene 20 minutes later may find only snow in the clouds as a result of glaciation. One airplane may go through some nice little wet imbedded cumulus cells, and the next airplane may miss them. Two airplanes 1000 feet apart in a stratus deck will often encounter different icing conditions. Two airplanes going through the same cloud at the same time at considerably different speeds may encounter widely different icing conditions due to differences in collection efficiency and ram temperature rise. Icing forecasting could be very much better than it is, but one of the difficulties it will always face is this variability.

If you pick up a load of ice, various nasty things are possible. We do our best, with avoidance techniques and equipment designed to prevent or remove ice, to avoid this. However, since it can never be ruled out, it is worthwhile to understand what ice can do to an airplane. A lot of data has been taken in icing tunnels and in flight testing with various airfoils and airplanes. Although the effects will vary from one airplane to another, it is possible to give some ball park numbers to point out the relative magnitude of the various problems and to possibly clear up some misconceptions. First of all, on the bright side, the weight of the ice does not, in itself, present a serious problem. Ice, as it collects on airplanes in flight, weighs about 50 pounds per cubic foot. If you consider how many boxes 1 foot on each side could be filled with ice taken from even a heavily iced airplane, you can see that the weight alone, while not totally insignificant, will not bring you down. Put another way, ice collection of 7 pounds per hour per foot of span is about all you would expect, even under very heavy icing conditions.

On the not so bright side, even small buildups of ice on airfoil leading edges can decrease the maximum lift coefficient by about 30 percent. Most of the damage to lift is done by the first accumulation, and increasing the size of the buildup then has less effect. The conclusion, unfortunately, is that even a small buildup will significantly reduce the angle of attack at which an airfoil will stall. This, of course, translates into increased stall speed and increased likelihood of an inadvertant stall in a turn or pullup maneuver. As an example, stall-speed data taken on an airplane with normal cruise configuration stalling speed of 69 knots showed that ⅛ inch of ice on the wing leading edge raised the stall speed to 80 knots, a 16 percent increase. A further 1.25 inches of ice raised the stall speed only another 4 knots.

Drag coefficient, on the other hand, starts to increase as ice starts to build and keeps right on increasing. An ice buildup more or less conforming to the airfoil shape will eventually double, or in severe cases triple, the drag of the airfoil. A horn-shaped ice formation with protrusions above and below the leading edge will triple the drag at about the same thickness that the more benign shape would double it, and may eventually increase the drag by a factor of 5. Other data which have been taken in terms of power, which may be more familiar than the engineering terms, have shown that the drag rise due to the horn-shaped ice buildup may raise the power required to maintain altitude in a given airplane by a factor of about 2.5. This, of course, may be more than the engine is capable of, particularly if propeller efficiency is also decreasing due to ice buildup.

Which brings us to the age-old controversy of which is better, keeping propellers clean or keeping airfoils clean. The propeller advocates lose the contest handily, according to the data. The power loss from iced airscrews is usually around 9 percent and very seldom more than 20 percent. This is quite a bit but still small by comparison to increases in power required to maintain altitude on the order of 250 percent because of ice on the wings. Prop unbalance due to possible uneven shedding of ice is another matter, but this seems to be a problem more likely to be due to malfunction of a propeller anti-icing system on one blade than to total lack of ice protection.

Handling characteristics of an airplane can also be adversely affected by ice accumulation. Roll control can be degraded by icing on the wings in front of the ailerons. Pitch control can deteriorate because of ice on the empennage leading edges and also because of ice on the wings affecting the downwash over the tail. If the position of the elevators in flight exposes the balances to ice accretion, enough ice may build to cause the elevators to bind against the horizontal stabilizer if sufficient clearance has not been provided.

Other components and systems of airplanes not designed for flight in icing conditions, or protected by suitable systems, can be affected in various ways. Windshields become opaque. Fuel vents exposed to the airstream can become plugged. Generators and avionics can overheat and fail due to blockage of their cooling-air scoops. Antennas can break off. Static systems not protected by heating or by their location can be affected. Other effects are imaginable, of course, but the ones mentioned are a few of the more common ones.

Obviously, the best way to deal with ice on the airframe is to avoid getting it, or to prevent or remove it with equipment provided for the purpose. However, once the ice is there, it behooves the pilot to operate with great care. Even a little ice justifies a 20 percent increase in approach speed because of the probable increase in stall speed. Do not go to a flap setting near the ground that you have not tried first with a little altitude beneath you. If at all possible, simulate the landing flare at altitude before it is necessary to do it for real. If disturbed airflow from the iced wings impinges on the tail during the flare, an abrupt pitch down might be the result. On landing, handle the airplane gingerly. Flare gently with no abrupt power reductions and fly it on. Consideration should be given to the length and condition of the run-away, since anything resembling normal short-field performance will not be possible.

This discussion would not be complete without some mention of the regulatory environment relating to flight into icing conditions. Airplanes presently come in three varieites in this regard. They are either certified for flight in icing conditions, prohibited from it, or neither. Since about 1972, the FAA has been alto-gether prohibiting newly manufactured airplanes from flight into known icing conditions unless and until they have been tested and shown capable of safely operating in the icing conditions specified for transport category airplanes in FAR 25. No amount of ice protection equipment installed on the airplane alters this prohibition until full certification has been obtained. Once certi-fied in this manner, however, the airplane receives virtual regula-tory carte blanche. The basic Part 91 contains no restrictions, and FAR 135 and Subpart D of FAR 91 permit airplanes with this icing certification to fly into known severe icing conditions. (Whatever that means, by the way. There is a large problem of definition here which has been discussed in Chapter 13.) Whether this all-or-nothing philosophy is wise is open to consid-erable debate, but it is all that exists at this writing.

Earlier airplanes were in many cases neither certified nor re-stricted. The only icing provision in the basic FAR 91 is the rule which requires a pilot to comply with the operating limitations of an airplane. This makes it illegal to fly an airplane with an icing prohibition into known icing. Otherwise, if the airplane is not prohibited from flight in icing and is not a large or turbojet pow-ered multiengine airplane covered by Subpart D, FAR 91 con-tains no icing operation rule. In addition, FAR 135 permits air-planes having certain listed items of ice protection equipment (and no prohibition against flight in icing conditions) to pene-trate light to moderate icing conditions (again, whatever that means) without icing certification. So, hire yourself a proverbial Philadelphia lawyer and figure it out. Good luck.

15 *Ice Protection Equipment*

There are a number of different types of ice protection equipment on the market. There's no real point to trying to give advice on what to buy, because ordinarily there's not much choice for any given airplane type. You buy what the manufacturer sells, unless you want to undertake the considerable time and even more considerable expense of certifying something else. Nonetheless, it's worth the effort to understand the capabilities and limitations of various types of equipment and what sort of performance can be expected from them.

For starters, the question to ask is whether or not the airplane you're concerned with is certified for flight into known icing. If it is, you can be reasonably sure that whatever equipment is on it will work pretty well *if properly maintained and operated in accordance with flight-manual instructions.* As previously mentioned, since about 1972 it has been FAA policy to placard airplanes against flight into known icing conditions, regardless of what equipment is installed, unless and until the manufacturer of the airplane has certified the equipment by showing that the airplane can safely be operated in the icing conditions specified for certification of transport (FAR 25) airplanes. Since about 1980, it is increasingly becoming FAA policy not even to allow the installation of ice protection equipment until it has been certified. However, that leaves a multitude of airplanes with ice protection equipment of some sort installed which have not been certified for flight into known icing. The differences between icing-certified and non-icing-certified airplanes are important and are sometimes subtle. On one model, for example, icing-certified airplanes have protected propeller governors and recessed fuel vents. Non-icing-certified airplanes don't, even though some have boots, heated windshields, and heated propeller boots. On another model, the boots on the non-icing-certified airplanes are smaller than those on the certified ones.

How, you might reasonably ask, was the equipment put on the non-icing-certified airplane then? Well, it's like this. It was done on what is known as a no-hazard basis, which means that the equipment was put on the airplane, and its presence was shown not to degrade performance or otherwise to compromise the safety of the airplane. That was the only requirement.

"Hmmmm," you might reasonably muse. "If it's not certified for icing, how do I know it will work if I need it?" The answer is, you don't. Now, airplane manufacturers are definitely not the ogres that certain ambulance chasers make them out to be, at least not in general. Some have done quite a lot of actual icing testing of equipment they offer even though they may not have certified it for one reason or another. Even if it works fine, however, it must be operated correctly to be effective. What follows is true in general, but note that it is not intended to supplant any flight manual instructions. If the book that comes with the airplane says something different, follow that book, not this one.

The preferred method of protecting airplane components against icing is heat, if sources of heat are available. Electric heat is usually used on small components, such as pitot tubes and propeller blades, and sometimes on windshields. Heating larger components, however really sucks up electric power and is avoided by manufacturers as much as possible. On turbine airplanes, heat for protection of leading edges and other large components, and sometimes also for the windshield, is often bled from the engine compressors. Other sources of heat have also been used occasionally. The C-123 military cargo airplane, for example, has a literal gas furnace which provides hot air for ice protection. Bleed air and electric elements are by far the most commonly used heat sources, however.

As a general rule, it is important to turn on heated ice protection systems *before* ice is encountered. There are several reasons for this. One is that the system may be very effective in preventing ice if the surface is warm and fully powered before the water hits it, but not so effective after ice has formed. Another is that ice which has formed on a surface acts as a thermal insulator, possibly creating hot spots which can damage electric heating elements which are turned on late. Furthermore, it is possible that if ice is allowed to form on some heated components and then removed, it will hit and damage something farther back on the airplane. In the case of propellers, turning the heat on late will fling chunks of ice into the fuselage and may cause vibration problems because of propeller unbalance due to uneven shedding. Good and commonly used rules are to turn on all systems protecting pitot tubes, static sources, and any angle of attack or stall warning sensors before entering any cloud and to turn on all other heated systems well before entering any cloud when the temperature is $+4°C$ or less. Any airplane prone to formation of carburetor ice should have a carburetor–air temperature gauge for each engine, and the carburetors should be heated to a safe temperature before entering any possible icing cloud. (Incidentally, remember that it's possible to ice a carburetor without even being in cloud and at temperatures well above freezing if the humidity is high. That was once very common wisdom, but perhaps it's not anymore since carburetors are becoming increasingly rare.)

On airplanes where heat for ice protection is not available, the next most common method of protection is pneumatic inflatable boots. On piston engine airplanes, the boots are usually sucked down by the vacuum side and inflated by the pressure side of the ship's vacuum pump(s). On turbine airplanes, bleed air is generally used for both inflation and deflation of the boots. You may well have heard a lot of derogatory comment on boots, as they are not too well thought of by many pilots. On icing-certified airplanes, they generally work pretty well. In some cases, however, a good bit of tinkering was done by the manufacturer to get them to work. It has sometimes been necessary to redesign the tube patterns, to adjust inflation pressures, and to play with inflation times, for example. To work properly, the boots must be kept clean and the system pressures and sequencing must be kept within manufacturer's specifications. On a twin-engine airplane with two vacuum pumps, there is redundancy for operation of the air-driven gyros but very possibly not for the boots, which may well not shed ice with only one pump driving them. Exactly how the plumbing is done and how the pressure is regulated will determine this on any given installation.

Inflatable boots, of course, do not prevent ice like heat. They break it off after it forms. It is necessary to let a certain amount of ice build up before operating the boots in order to shed effectively. A specific airplane's flight manual should be consulted for instructions on when to cycle the boots. In the absence of such instructions, about one-half inch is as good a ball park thickness to use as any. Much less than that and small pieces of ice are likely to stick to the tubes as they inflate, giving further ice a safe place to form and not be broken. Too much more than that and the ice may not break. Unfortunately, it is impossible to be too specific on this subject, since there will be more ice on the empennage boots than on the wing boots anyway (as a result of the higher collection efficiency of the smaller components) and since the only common method of determining the ice thickness is the old eyeball. It is obvious, I would assume, that a good light is required to inspect the boots for ice accumulation at night. Even if the airplane has ice-inspection lights installed, as is generally the case for airplanes certified for flight in icing conditions, a good flashlight should be carried for backup. As the old joke goes, if you turn on your light and don't like what you see, you can always turn it off.

After heat and boots, the most common ice protection systems are fluid systems. They usually use an alcohol or glycol mixture to either prevent or remove ice. Some common applications are slinger rings on propeller hubs for blade anti-icing, buttons for nose and radome protection, and distribution rakes of some sort for windshield clearing. There is a British-made system which uses a porous leading edge to ooze out a glycol coating on airfoil surfaces which is certified on several airplanes and is performing very well in some NASA icing wind-tunnel testing in progress at this writing. Properly designed fluid systems are very effective and have the advantages of simplicity, low maintenance, and low-power requirement. Their obvious disadvantage is the requirement to carry expendable fluid, which is heavy and which sooner or later runs out. They can also be messy and have the potential of ruining clothes for the unwary passenger who brushes against them. If you have a certified fluid system, about all that it's necessary to say is to operate it according to the manufacturer's instructions and keep the reservoir topped.

The ultimate panacea for icing, of course, is some kind of material or goo which can be installed or buttered on an airplane and on which water will not freeze. This subject has been studied extensively by NASA and others, and nothing practical has been found to my knowledge. Some substances have been found marginally effective as release agents in combination with other systems, and some may work for a very short time, but that's about it. My own personal experience is that ordinary car wax works about as well as anything on boots, and it probably just keeps them clean. My own bet, as a devoted Murphy's law believer, is that when science finds a substance which water won't stick to, the next discovery will be that the stuff won't stick to airplanes either.

Finally, the Russians have a U.S. patent on an electroimpulse deicing system, and there have been a few British and French experiments with it. From what I have been able to find out, which is not a whole lot, the system was originally invented to de-ice ship's masts, in which application it may work pretty well. The Russians then put it on a few large airplanes. It apparently charges up to a high energy, then discharges with a mighty ka-*pow*, fetching the leading edges a clout which breaks off the ice and very possibly anything else not firmly attached. I don't think I'll hold my breath waiting for any U.S. manufacturer to use that particular system.

16 *Icing Certification*

The importance of icing certification has been mentioned a number of times in the last two chapters as an assurance that an airplane is adequately protected against the problems that icing can create. It may be instructive for an appreciation of this to take a look at what is done to certify an airplane for flight into known icing. Historically, modern icing certification goes back to the work done in the late forties by NACA. The most important single reference is probably NACA Technical Note 1855, published in 1949, which established a number of icing envelopes for use in aircraft design. Two of these envelopes, called "continuous maximum" and "intermittent maximum," were adopted as part of the certification standards and are now found in Appendix C of FAR Part 25.

Although icing certification is referred to in other parts of the FAR, they all finally refer to these envelopes, which are shown in Figures 16-1 and 16-2. The liquid-water content is grams of liquid water (not snow or ice crystals) per cubic meter of air. The drop size is given in microns. There are 1000 microns in a millimeter. The sizes given are typical of icing clouds. A small drizzle drop is around 100 microns in diameter. It is important to note that these envelopes are science fiction in much the same sense that the standard atmosphere is science fiction. You can't go out

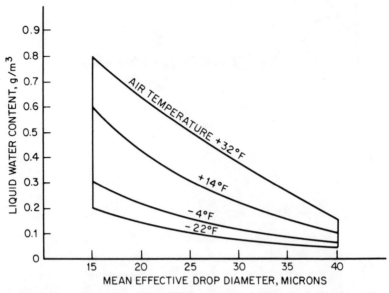

FIGURE 16-1 *Continuous maximum (stratiform clouds) atmospheric icing conditions. Pressure altitude range: sea level–22,000 feet; maximum vertical extent: 6500 feet; horizontal extent: standard distance of 17.4 nautical miles.*

and find them in the atmosphere. Finding an icing condition at random which falls within the envelopes is not too difficult but to find any given point, especially any point near the corners of the envelopes, is well-nigh impossible. What these envelopes are is an engineering standard for aircraft designers, and what they are really saying is that if the designer protects the airplane against all icing conditions defined by these envelopes, the airplane will be able to operate safely in the real icing conditions which it will encounter in service. Whether this was really true could have been on pretty shaky ground when these envelopes were first developed, but they have proven to work very well over the years and they are still in use. Note also that these envelopes do not address any definition of whether an icing condition is light, moderate, or severe.

Very well then, the airplane designer takes these envelopes as given and considers how these theoretical icing conditions could

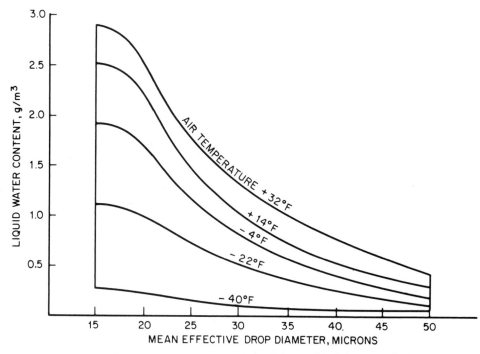

FIGURE 16-2 *Intermittent maximum (cumuliform clouds) atmospheric icing conditions. Pressure altitude range: 4000–22,000 feet; horizontal extent: standard distance of 2.6 nautical miles.*

affect an airplane being designed. Adequate levels of performance, stability, control, engine operation, cockpit visibility, and systems operation must be maintained in these icing conditions. Decisions are made on what parts of the airplane will require some sort of protection and what parts can be left unprotected. For those areas that require protection, a system is designed to do the job. Where heat is used, the designer uses the FAR 25 envelopes to calculate how much water could be collected on the heated section and makes sure that a sufficient source of heat, usually from either hot air or electric elements, is available. Where boots are used, an analysis is performed using the specified drop sizes to make sure that the water will hit the boots and not go around them. When fluid is used, the designer makes sure that sufficient fluid flow is provided and that the fluid will flow over all the required area at any angle of attack and configuration which will be used in icing conditions. Any other requirements unique to a given airplane are considered, and whatever protective systems may be required are designed. The total ice protection package is then installed on a prototype airplane, and flight testing begins.

The first thing which is usually done is to verify that the equipment itself constitutes no hazard. If large electric heating elements are used on an airplane with two generators, for example, it must be proven that the elements don't suck up so much juice that if one generator fails when they are turned on, the other one will quit or be wrenched off its pad as a result of the sudden torque increase—thereby bidding sayonara to the whole electric system. Boots, if they are used, must be cycled at high speed and at or near stall speed to make sure that they don't cause any buffet or stall-characteristics problems. Heat provided for windshield protection must not melt the windshield. Possible electrical interferences are looked for and corrected if found. If bleed air from turbine engines is used, tests are run to verify that the bleed-air extraction effects are known and do not cause any problems to the engines themselves. As was previously mentioned, this sort of no-hazard testing was all that was required, and in some cases all that was done, on older airplanes which have some ice protection equipment installed but which are not certified for flight into known icing. This, as you can plainly see, tells you nothing about how well the equipment will protect you from ice.

After everyone is satisfied with the no-hazard tests, the next usual step is testing in dry air. The temperatures of heated sections are measured under a number of flight conditions to be sure that the heat is adequate. Boot inflation and deflation pressures and times are measured. Fluid-coverage patterns are looked at. If any potentially critical areas, such as the leading edges of vertical or horizontal tail surfaces, are left unprotected, simulated ice shapes are installed on these surfaces and the airplane handling characteristics are investigated to be sure that there is no unacceptable degradation.

When everything that can be done in dry air passes muster, it's time to put water on the airplane. Nowadays this is frequently done first with a tanker spraying water. This makes it possible to check the effectiveness of the systems on the test aircraft one at a time and lets the test crews choose their own conditions and locations. The safety advantages to this method are obvious. It also makes it easy to take pictures from a chase airplane of such things as wheel-well doors which can't be seen from inside the test airplane. In addition, it is possible by using a tanker and chase airplane to observe the trajectories of chunks of ice shed from unprotected areas of the test airplane when it descends into warmer air and to be sure that these won't do such nice things as go into a turbine engine inlet or shear off an antenna. Tanker testing is an excellent method of making sure that pneumatic boots will, in fact, break ice so that it will shed from the protected surface. If they don't break it or if the ice breaks in such a way that cuffs are left over the surface and the ice stays with the airplane, it is easily possible to make necessary changes and check the results with another tanker test.

All the above tests are usually done by the manufacturer before the FAA is called in for certification testing. The manufacturer may also do some tests in natural icing or may omit the tanker tests if the systems in question are very similar to systems already used on other airplanes or if tanker tests are otherwise felt to be unnecessary. In any case, when the manufacturer feels that the airplane is ready for icing certification, the FAA is notified. Their engineering and flight-test people review the manufacturer's tests and issue a document called a "Type Inspection Authorization," which specifies the tests which they feel are necessary to demonstrate that the ice protection systems meet the requirements. This usually involves repeating the dry-air and tanker tests which the manufacturer showed were critical: e.g., warm-air tests are not repeated if tests done in cold air are sufficient to assure that the systems will also work in warm air. It always involves tests in natural ice. Enough icing must be encountered to demonstrate that all the ice protection systems work to protect the entire airplane. Various configurations and speeds are checked, and handling qualities are generally checked with ice accumulations on all the unprotected areas in at least the approach and landing configurations (see Figure 16-3). When all these tests are satisfactorily completed, the airplane is granted FAA certification for flight into known icing conditions.

In a previous chapter, it was mentioned that airplanes which have been certified to the FAR 25 envelopes, which is the only way to certify them at this writing, are under no restrictions in regard to the intensity of icing conditions which they can be flown in under the operating rules. Nowadays, if a new airplane is not so certified, it cannot be flown into any known icing condition, regardless of severity, because it gets an operating limitation which prohibits it. There is quite a bit of debate going on about

FIGURE 16-3 *Horn-shaped ice collected on an unprotected wingtip outboard of the boot during an icing certification test. (Cessna Aircraft Company.)*

whether this is wise. The manufacturer presently has to choose either an expensive certification program that results in an airplane certified for flight into conditions in which the manufacturer may not wish to have it flown or a prohibition against any operation in known or forecast icing conditions whatever. In other words, in order to allow an airplane to penetrate 2000 feet of stratus with known bases and tops in which icing is forecast, the manufacturer has to provide protection for all icing conditions which could ever be encountered by a 727 or DC-10. This seems unreasonable to me, but it's the way it is today.

17 *Nonconvective Turbulence and Wind Shear*

It was January 1964, according to one of my old logbooks. I was a young charter pilot and flight instructor, and on this particular evening I had been hired by a local company to fly a Cessna 210 with three of their people from Erie, Pennsylvania, to a small town southwest of Cleveland, Ohio, with a well-lighted airport. As you have probably guessed, since I'm taking the trouble of spinning this yarn, it turned out to be a far more entertaining evening than any of us had bargained for.

It had been a beautiful day, a rarity for that time of year in the Great Lakes and had become an equally beautiful night with no hint of trouble in the forecasts when we took off from Erie and headed southwest down the lakeshore. Imagine my surprise when I soon found myself bucking something like a 40-knot headwind. About the time we passed Cleveland, cruising at a fairly low altitude (the winds get lighter down lower—don't they?), it got so rough I'd swear I had white caps in my coffee. I was concerned enough about the turbulence that I slowed the airplane down. I told my passengers, who thankfully were veteran flyers and at least were not getting airsick, that if this kept up we might not be able to land, but since we had plenty of fuel to get home, we would go take a look. We duly arrived overhead of the destination airport, and the lighted wind sock indicated very little surface wind. I warily set up an approach, and as I descended into a downwind pattern entry, the turbulence died. You're not fooling *me,* I thought, as I lowered the gear but held the flaps up waiting for the next clout. It didn't come, however, as I turned base and then final. In the usual clarity of hindsight, of course, the drift-correction angle I was holding, combined with what the wind sock was telling me about the surface wind, should have tipped me off. It didn't, of course, so I slowed the airplane down, fat, dumb, and happy, and like a true P.T. Barnum sucker, I extended full flaps. *Pow!* Everything turned to worms.

I felt the sink when we hit the shear and slammed the throttle to the wall. The only thing visible in the landing lights was a big tree off the left side of the runway, which we were pointed directly at. I was vaguely aware of a pain in my right leg, but I ignored it as I slammed in all the right aileron and rudder I had. In less time than it takes to tell it, I realized that we were back under control over the runway and had almost settled onto it in spite of the power I was carrying. All I had to do to land was close the throttle, so I did it, and the next thing I knew we were rolling out on the runway as pretty as you please. That's when I discovered that the passenger beside me had his left hand dug into my right leg hard enough to leave me with a set of black and blue fingerprints, thumb to pinky, that were to last me the better part of a week.

About 3 hours later, after the passengers came back and, to my considerable surprise, got back into the airplane, we took off for the return trip. Being snakebit and expecting the worst, I held the airplane just off the ground building up speed for the full length of the runway and then pulled up into the darkness. Nothing. Absolutely zilch. Not a ripple. The flight home was just as uneventful as the flight over was supposed to have been. The strong wind and turbulence were gone. I couldn't believe it. It was years before I finally understood and was able to explain what probably went on in the atmosphere that night. I'll let you in on it in just a bit.

First, though, let's give some thought to the threats that turbulence can pose to an airplane. They basically fall into three areas: structural integrity, performance, and controllability. If an airplane is prudently flown in accordance with the manufacturer's recommended turbulence-penetration speeds, thunderstorm turbulence and possibly turbulence caused by a severe mountain-wave rotor are the only kinds that will ever threaten structural integrity directly. In the episode just described, however, my airplane was pushed to the limit of both control and performance by turbulence totally unrelated to mountains or thunderstorms. In order to have some appreciation for what we can expect from an airplane in turbulence, it may be instructive to take a brief look at the relevant certification criteria.

There are a lot of airplanes flying, and quite a few still in production, which were certified to the old Civil Air Regulations Part 3 or Part 4, under which the gust requirements were considerably less than they are nowadays. However, we will limit the present discussion to the regulations current as of this writing, which are FAR 23 for light airplanes (which also covers most general aviation turboprops and a few jets) and FAR 25 for transport category airplanes (which also covers the majority of business jets). Both regulations require that airplanes be designed to withstand without structural damage symmetrical vertical gusts, both upward and downward, of 3000 feet per minute at design cruising speed up to an altitude of 20,000 feet. Above that altitude, the required gust velocity is slowly reduced to 1500 feet per minute at an altitude of 50,000 feet. Understanding what the term "design cruising speed" means requires wading through some legalese, but to a pilot it basically means maximum operating limit speed (for an airplane which is both speed- and Mach-limited) or the speed which separates the green arc from the yellow arc on the airspeed indicator (for an airplane with a never-exceed speed and without a defined Mach limit). FAR 25, the transport regulation, then goes quite a bit further. It also requires that an airplane be able to withstand a vertical gust of almost 4000 feet per minute below 20,000 feet at a slower speed to be used for turbulence penetration, and it has some unsymmetrical vertical-gust and lateral-gust requirements. I suppose that one could philosophize endlessly upon whether these additional requirements make transport category airplanes "safer" than nontransport or whether they just account for the fact that FAR 25 governs larger and much more flexible airplanes than FAR 23. Structural design of an airplane also requires consideration of maneuvering loads, which, particularly in the case of an FAR 23 airplane, may well be higher than the gust loads anyway. I don't see any point to entering into that discussion here, since the intention is just to give some appreciation of the magnitudes involved and not to try to teach an engineering course. An airplane certified to either regulation is sufficiently stout if prudently flown, but there ain't no such thing as an "all-weather" airplane in either case.

In the area of performance, neither regulation has any requirements specifically directed toward operation in turbulence. There has been ample evidence in the last few years that the climb performance of airplanes certified under either regulation can be overwhelmed by turbulence and wind shears, particularly when they are caught in other than the optimum-climb configuration and at low altitude. I would personally be at a loss to propose any regulatory fix to this. The only answer I know of is to recognize and avoid potentially catastrophic weather. That's what this book is about, after all.

The requirements for controllability are a somewhat different story. Both FAR 23 and FAR 25 recognize the possibility of so-called turbulence upsets for airplanes flying at high altitude and high speed, and they both require the same sort of testing to show that a margin of control and speed exists which is sufficient to recover from any reasonably foreseeable departure from trimmed flight that might be induced by turbulence. This is well and good. It seems to me, however, that the lateral-control requirements of both regulations in situations in which the airplane is close to the ground leave a little to be desired. Both regulations essentially require that airplanes be controllable in crosswinds up to 20 percent of the landing-configuration stalling speed, which I suppose could be thought of as a turbulence requirement of sorts. In addition, FAR 25 at least recognizes that turbulence problems exist in requiring that "roll response must allow normal maneuvers (such as recovery from upsets produced by gusts . . .)." However, at this point it leaves both the designer and the test pilot hanging in regard to what that means. FAR 23, on the other hand, does not mention low-level turbulence but requires that airplanes weighing 6000 pounds or less be able to roll from a 30° bank in one direction through a 30° bank in the other direction in 5 seconds in the takeoff configuration or in 4 seconds in the approach configuration. More time is allowed for heavier airplanes according to formulas which give a 12,000-pound airplane almost 10 seconds to complete the 60° roll in the takeoff configuration and almost 7 seconds in the approach configuration. Most light airplanes are capable of roll rates well in excess of this, which is a good thing. Five seconds would be bad enough, and ten seconds would be downright ponderous for a 60° roll. NASA released a report in 1966 on the handling qualities of seven commonly available light airplanes which they evaluated. While the comments were less than glowing in some respects, they did conclude that all the airplanes would meet what would be the appropriate military specification for roll rate, which is far in excess of the FAR requirement. One of the airplanes, however, was called sluggish in roll, and this was related to its roll acceleration, or the time required after aileron application to achieve a given rate of roll. Neither FAR 23 nor FAR 25 have any requirement for roll acceleration, which may well be the most important factor in recovering from a sudden turbulence-induced roll.

So there you are. I want to make it clear that the foregoing is not intended as an indictment of either the certification rules or any particular airplane. Although some would certainly disagree, I don't think the safety records are bad enough to deserve that. My own rather lengthy experience has been that in most severe weather accidents of which I have personal knowledge, the weather was obviously awful and no regulatory change would have been likely to help much. Tweren't a fit day out for man or beast. A crusade to improve weather service, and the level of knowledge of weather effects in general, would probably be more effective than one to overhaul the certification regulations. I do think that the requirements on roll control should be improved, however.

Now then, so much for the airplanes and back to the weather. We've already said quite a bit about thunderstorm turbulence and wind shear in previous chapters and discussed the effects of the several possible cases of wind shear encounters. Since most encounters with high-altitude clear-air turbulence are just obnoxious, not dangerous, and since the dangerous kinds which can cause upsets at high altitude have been pretty well written up elsewhere, let's concentrate on the low-level turbulence and shears which can tend toward unorthodox encounters with terra firma. There are several things skulking about in the atmosphere besides thunderstorms which can cause potentially hazardous low-level shears.

Let's go back to that night that tried to do me in back in northern Ohio. What probably happened is this: Ordinarily, as a parcel of air is tooling along with the wind in a layer near the ground, it is acted upon by three forces. The force which moves it along is provided by the good old pressure gradient, which is too much beaten to death in every elementary weather course for pilots to need any further publicity here. The force which keeps it from going straight into the low-pressure center is the Coriolis force, which acts at right angles to the direction of motion and tries to make the parcel turn to the right (in the northern hemisphere, for the benefit of the mathematicians). The third force, which retards the parcel's motion, is friction with the ground. Ordinarily, these forces are in equilibrium and what you see is a vertical wind profile in which the wind velocity is more or less constant at any given level. As you ascend, the wind speed increases (the surface frictional effect is less) and the air turns to the right (more Coriolis force due to the higher speed). These changes are gradual and uniform. Now, however, let's imagine that the sun goes down and a low-level inversion forms over fairly flat terrain as the ground cools off. The frictional force with the ground acting on the air just above the inversion goes away, and the air parcels in this layer are now in much the same situation that a heavy truck inching its way down a steep hill would be in right after it hit glare ice. The forces are no longer in balance, and Newton's laws say that an object acted upon by an unbalanced force accelerates. That's exactly what happens, and the result is a ribbon of high-speed wind where none would ordinarily be expected. Some National Weather Service material on low-level wind shear forecasting which they were kind enough to share with me calls this a low-level jet (not to be confused with the low-level jet mentioned in connection with severe thunderstorms, which is altogether different) and defines it as a wind-speed profile of 0 to 8 knots at the surface, increasing with height to more than 25 knots between 650 and 1500 feet and then decreasing in speed with height. The reference also states that forecasting the formation of low-level jets of this type is a problem. This is true. They seem to be more common in the midwest than elsewhere but could probably occur anywhere in flat terrain where a strong inversion can form near the surface at sunset. They occurred numerous times while I was forecasting weather for the Air Force in eastern New Mexico several years after my Ohio surprise. I found that they seemed to be most likely when the altimeter setting was low, about 29.70 or less, and that they would sometimes tip you off by tickling the leaves in the tops of the trees just after daylight. I also found that, with the facilities of a weather station at my disposal, it was very easy to confirm my suspicions that a low-level jet might be overhead. All that was necessary was to launch a balloon and watch it. There was no need to get out the instruments and track it. If the jet was there, the balloon would rise pretty much vertically for a minute or so and then take off like a scalded cat for west Texas or Oklahoma.

So how does a pilot deal with a low-level jet? The first part of the battle, of course, is just to realize that these things exist. If you encounter unexpectedly strong winds at low altitudes between the time a low-level inversion forms around sunset and the time it breaks a bit after sunrise, you can suspect that you are in one. As we've seen, the way out of the wind is up, not down. If climbing a couple thousand feet brings the wind down to something much closer to what you expected, you probably have a low-level jet underneath you. You may find feisty turbulence around any of its edges. National Weather Service guidelines give 6 knots per 1000 feet vertically and 40 knots per 150 miles horizontally as values of wind change indicative of shear critical for turbulence, and both of these are easily exceeded by a low-level jet. For takeoffs with a calm or nearly calm surface wind, it is an excellent idea any time a low-level inversion may be present to take off in the direction that would be upwind according to the existing weather situation if there were a wind. The wind is likely to increase through the inversion anyway, and if a low-level jet is there, the pilot who takes off in the wrong direction is all set up for what we earlier called a Case Three wind shear encounter when we were discussing thunderstorms, i.e., climbing into a suddenly increasing tailwind. This can be a particular gotcha for the crop duster who bores off at first light and pulls a heavily loaded agplane up into whatever's there. Pilots have been told about the presumed hazards of downwind turns for years and cautioned to carry some extra speed and be sure to use the airspeed indicator rather than reference to airplane motion over the ground when making such turns. I'm sure that's good advice, but the fact remains that a flying airplane doesn't know aerodynamically which way it's going with respect to a steady wind. It sure knows when it's pulled up into a tail-on wind shear, though. I suppose we'll never know how much of a part a wind shear factor of this sort has played in accidents in which an airplane fell out of a low-level turn in some manner, but I'll bet it's been plenty. If you're landing out of a low-level jet, like I was that night, be smarter than I was. You have a choice of which way to take the shear. Consider your options in terms of runways available and the capabilities of the equipment you're flying and take it as a crosswind or as a Case Two encounter (in which you descend out of a tailwind, referring again to the thunderstorm discussions), not as a Case One (in which you descend out of a headwind) like mine primarily was. Or be smarter still, and don't do it. You may be able to outsmart your performance problems by taking the shear in the best direction, but the control power of the airplane and your reactions still have to be up to dealing with the turbulence.

Low-level jets of this sort are not the only cause of potentially hazardous nonconvective low-level wind shears and turbulence, although they are certainly one of the sneakiest. Any low-level inversion has the potential for wind shear, since it basically separates the wind field above the inversion from the wind field below it. Both cold fronts and warm fronts have this potential. National Weather Service guidelines are that any front producing a temperature change of 10°F or more in 50 nautical miles *and* with a wind of 40 knots or more at 2000 feet above the ground can produce significant low-level wind shear. Mr. Daniel F. Sowa of Northwest Orient Airlines, who has probably done more than any other meteorologist to get wind shear and turbulence knowledge into the hands of pilots where it'll do some good, has published criteria of *either* 10°F across the front *or* a speed of movement of the front of 30 knots or more as indicative of low-level shear. He has also published information to the effect that a larger temperature difference across the front can create wind shear with slower frontal movements and will also increase the likelihood that the shear zone will be turbulent. Fast-moving cold fronts can easily create shears in the neighborhood of 40 knots in 200 feet, with both the along-track and crosswind component along a typical approach path possibly changing somewhere in the neighborhood of 20 knots. If the direction of the instrument approach in use is more or less into the prevailing surface wind, as is usually the case, cold frontal shear will usually be of the increasing headwind type for airplanes landing out of this approach. The shear hazard will be present *after* the surface position of the front passes the airport, and if the front is fast moving will only be at a low altitude for a short time due to the steep cold frontal slope. While a suddenly increasing headwind type of shear certainly beats a suddenly increasing tailwind (or dying headwind) at low altitude, it can still create a lot of problems. The indicated airspeed will increase suddenly, tending to snooker the pilot into pulling off a handful of power. This will tend to increase the rate of descent after the airplane stabilizes, however, and the increase in headwind will decrease the groundspeed. If the power and groundspeed before the wind shear encounter were holding the glide slope, both the change in groundspeed and the power decrease will tend to drop the airplane in short. The more inertia the airplane has and the longer the engine response to throttle commands takes, the worse this sort of problem will be. An airplane with enough power and energy to be easily capable of a go-around, and a pilot willing and ready to do it, is the best recipe for avoiding the stumps and alligators in this situation. You'll have a lot better idea of what to expect the second time around anyway if the first one doesn't look good.

Wind shear hazards from warm fronts will be present *before* the frontal surface at the ground passes the airport, and can usually hang around at low altitude a good bit longer than shear hazards from cold fronts because of the warm front's lower frontal slope. Surface winds ahead of a warm front are often fairly light, which makes it difficult to say what direction the approach in use might take or to generalize on the shear. In any case, it's always up to the pilot to evaluate the situation and compare whatever the best estimates of the wind above the inversion are with the wind near the surface to decide what sort of wind shear encounter is likely. Again, as we found before in the case of severe thunderstorm development, a warm front is not necessarily the pussycat (relative to a cold front) that it is frequently made out to be.

A fairly strong wind near the surface tends to oppose the formation of a low-level inversion by keeping the air in the lower levels pretty well mixed. A very cold surface, relative to the air over it, will act in the opposite direction. An example is ground which is covered with fresh, cold snow. If an inversion forms over such a cold surface where there has been a fairly strong wind and if there has been no other reason for the wind to have died (i.e., weakened pressure gradient in the area), then it hasn't. It's still there above the inversion, possibly strengthened, and now there's a wind shear. Also, if such a cold-surface-induced inversion forms in a calm region and if the pressure gradient increases due to the approach of a low-pressure system, the inversion may not break and again a wind shear is possible. In mountainous areas, a strong downslope wind may occur above such an inversion. The point here is not to compile a list of all possible low-level wind shear causes, however unlikely, which no one would ever remember anyway. The point is just to reemphasize the fact that practically any low-level inversion, whatever the cause, makes a low-level wind shear possible. When such an inversion is likely, the pilot should look for any reasons why a change of wind might exist through the inversion, and act accordingly.

The subject of wind shear certainly can't be said to have been neglected in the last few years. The FAA published a bibliography in 1977 (FAA-RD-76-114, available from the National Technical Information Service, Springfield, VA 22151) which lists no less than 216 documents on wind shear, the vast majority of which have been written since 1970. It's a great place to start for anyone who wants to study the subject further. A lot of work is going on to better understand and identify wind shear problems. Computer programs are being developed to allow pilots to experience realistic wind shear problems in simulators. This is all good and will help to make everyone's flying just a little safer, if it doesn't wind up pigeonholed on a government shelf where no one who could use the information would ever look. If this book has served the purpose of getting some of the word out on this and the other subjects we've discussed, and if it helps even one pilot to make a correct and safe weather decision sometime, it will have been worth doing. Cheers.

Index

About the author:

Dennis W. Newton is Chief Engineering Test Pilot for LearFan Corporation and holds licenses as Airline Transport Pilot, Flight Instructor, and Ground Instructor. He has been a flight test engineer, test pilot, and flight operations manager for Cessna Aircraft Company. He was chief research pilot of the Pennsylvania State University department of meteorology, has published NASA sponsored technical papers on icing, and has served as a weather forecaster in the U.S. Air Force. He is a member of the Society of Experimental Test Pilots and the American Institute of Aeronautics and Astronautics.

AUG 2 0 1984			
MAR 1 1986			
NOV 1 3 1986			
DEC 0 1 1986			
DEC 15 1986			
JAN 0 8 1992			